10 0447977 5

KU-279-713

WITHDRAWN FROM THE LIBRARY

UNIVERSITY OF NOTTINGHAM

ONE WEEK LOAN

UNIVERSITY LIBRARY

- 8 NOV 2005

OWL GGL 08

UNIVERSITY LIBRARY

1 2 DEC 2006

OWL GGL 18

UNIVERSITY LIBRARY

1 5 JAN 2007

OWL GGL 03

UNIVERSITY LIBRARY

- 9 DEC 2013

GGL 27

A fine will be charged if not returned by the date stamped above

CHAPMAN & HALL/CRC MATHEMATICS

Full information on the complete range of Chapman & Hall/CRC Mathematics books is available from the publishers.

SETS, FUNCTIONS, AND LOGIC

AN INTRODUCTION TO ABSTRACT MATHEMATICS

THIRD EDITION

KEITH DEVLIN

CHAPMAN & HALL/CRC

A CRC Press Company

Boca Raton London New York Washington, D.C.

Library of Congress Cataloging-in-Publication Data

Devlin, Keith J.
 Sets, functions, and logic : an introduction to abstract mathematics /
Keith Devlin.-- 3rd ed.
 p. cm. -- (Chapman & Hall mathematics)
Includes bibliographical references.
 ISBN 1-58488-449-5
 1. Mathematics. I. Title. II. Series.
QA37.3.D49 2003
510--dc22 2003055714

This book contains information obtained from authentic and highly regarded sources. Reprinted material is quoted with permission, and sources are indicated. A wide variety of references are listed. Reasonable efforts have been made to publish reliable data and information, but the author and the publisher cannot assume responsibility for the validity of all materials or for the consequences of their use.

Neither this book nor any part may be reproduced or transmitted in any form or by any means, electronic or mechanical, including photocopying, microfilming, and recording, or by any information storage or retrieval system, without prior permission in writing from the publisher.

The consent of CRC Press LLC does not extend to copying for general distribution, for promotion, for creating new works, or for resale. Specific permission must be obtained in writing from CRC Press LLC for such copying.

Direct all inquiries to CRC Press LLC, 2000 N.W. Corporate Blvd., Boca Raton, Florida 33431.

Trademark Notice: Product or corporate names may be trademarks or registered trademarks, and are used only for identification and explanation, without intent to infringe.

Visit the CRC Press Web site at www.crcpress.com

© 2004 by Chapman & Hall/CRC

No claim to original U.S. Government works
International Standard Book Number 1-58488-449-5
Library of Congress Card Number 2003055714
Printed in the United States of America 1 2 3 4 5 6 7 8 9 0
Printed on acid-free paper

UNIVERSITY LIBRARY
NOTTINGHAM

Contents

Preface

This book is designed to help the university student make the difficult transition from calculus to university-level pure mathematics. A quick glance at the contents page will indicate to any instructor what the book sets out to do.

I first starting teaching this material in the late 1970s at the University of Lancaster in England. I wrote the first version of this book, published in 1981, to accompany a 6-week course I developed there. Teaching the same kind of material a decade later in the U.S., I looked at the original version, found it wanting, and completely rewrote it. I changed almost everything: I reordered the topics, changed the treatment of each topic, and increased the number of exercises. That second edition was published in 1992.

One decision I faced when I wrote the second edition was the size of the book. The original edition was deliberately written as a "little book" that would be both unintimidating and cheap. At the time, there were no competing books on the market (at least in the U.K.), and as a result the book did quite well. By the time I came to write the second edition, there was a whole slew of books (in the U.S.) targeted at the same students. All of these books were far more substantial than mine and, in keeping with the trend in textbook publishing, getting larger all the time in a furious struggle to be all things to all students and to include everything that appeared in any competing book. I looked at those books and decided firmly — I had to be firm because my publisher had other ideas — to stick with my original plan: Despite the many changes that I felt were necessary, the second edition would remain a "little book".

Now another decade has passed and the time has come to rewrite the book once again. Not because the core material has changed; that is the same now as it was when I myself was a student in the 1960s. What has changed is the background that the beginning student brings to his or her study and the environment in which the study is carried out. This new version of the book tries to take those changes into account. But I still believe that for this material, small is better.

Students Start Here

If you are a mathematics instructor, this book will tell you absolutely nothing you do not already know. Assuming you have read the Preface (which was written for instructors, under the assumption that students never read prefaces but instructors usually do), you really do not need to read any further in this book than this paragraph.

Those two sentences are the last ones in the entire book directed at instructors. Everything else (including this paragraph) is written for the beginning student of post-calculus university-level mathematics ("you", from now on).

You have completed courses on differential and integral calculus. Maybe you aced those courses; or maybe you got through only after a struggle. Either way, now you are trying to make the transition to what comes next. Chances are, regardless of your performance in mathematics up until now, you are going to find the next step unfamiliar and challenging. This book won't make it easy. No book can do that. Anyone who claims otherwise is trying to sell you a book — probably a thick, expensive one at that.[1] But after guiding many generations of young people through precisely this difficult transition,[2] I think I can help.

The main distinction between most high-school mathematics and post-calculus university math lies in the degree of rigor and abstraction required at the university level. In general, you (a student embarking on post-calculus university mathematics) will have had little or no prior experience of wholly rigorous definitions and proofs. The result is that, although you may be competent to handle quite difficult problems in calculus, you are likely to find yourself totally lost when presented with a rigorous definition of limits and derivatives — the fundamental mathematical ideas that lie behind calculus.

In effect, what you need in order to progress further in mathematics is to acquire mastery of what is virtually an entire new language ("the language of mathematics") and to adopt an entirely new mode of thinking ("mathematical thinking"). Throughout my many years teaching university mathematics,[3] I have met very few students who came through this process without a great deal of difficulty. (I certainly didn't when I was a student.) This book is intended to assist you in making this transition.

Chances are that almost everything you find in this book will be new to you and will probably seem very strange. Indeed, you may feel that it is not "mathematics" at all. Be patient. Given time and a fair amount of hard work, this stage will pass.

[1] See the Preface.

[2] See the Preface again.

[3] See the Preface yet again.

Do not try to rush through any part of the book, even if at first glance a particular section looks easy. This entire book consists of *basic* material required elsewhere (indeed practically *everywhere*) in post-calculus mathematics. Everything you will find in this book is included because it generally causes problems for the beginner. (Trust me on this. Hey, I got you to look at the Preface, didn't I, and how often have you done *that* with a math textbook?)

Part of the reason you are likely to find this material difficult is that it will seem unmotivated. It is bound to: the sole motivation is to provide you with the foundation on which to build the mathematics that comes later — mathematics that you do not yet know about!

So take it steadily, and try to *understand* the new concepts as you meet them. There is little in the way of new facts to learn, but a great deal to comprehend! (The actual facts contained in this book could be listed on three or four pages of notes.) And try the exercises — as many of them as possible. They are included for a purpose: to aid your understanding.

Discuss any difficulties that arise with your colleagues and with your instructor. Do not give up. Students all around the world managed it last year. Likewise the previous year, and the year before that. So did I. So will you!

Keith Devlin
Stanford University

Chapter 1

What Is Mathematics and What Does It Do for Us?

You might think these are odd questions to begin with. After all, won't you, my intended reader, already know what mathematics is? And doesn't everyone know that it's important to be able to "do math" — which presumably implies that they think it is useful stuff?

Maybe yes. But in my experience, given the way mathematics is often taught in schools — and in some universities, come to that — many students are never given a "big picture" view of the subject. I've long lost count of the number of adults I've met, many years after they have graduated with degrees in such mathematically rich subjects as engineering, physics, computer science, or even mathematics itself, who have told me that they went through their entire education without ever gaining a good overview of what constitutes modern mathematics. Only later in life do they start to catch a glimpse of the true nature of mathematics and the extent of its pervasive influence in modern-day society.

It's not hard to understand why this is the case. Most of the mathematics that underpins present-day science and technology is no more than 300 or 400 years old, in many cases less than a century old. Yet the typical high school curriculum covers mathematics that is at least 500 years old — in fact, much of what is taught is over 2,000 years old. It's as if our literature courses gave students Homer and Chaucer but never mentioned Shakespeare, Dickens, or Proust.

1.1 It's Not Just Numbers

Most people think that mathematics is the study of numbers. That description of mathematics ceased to be accurate about 2,500 years ago! Anyone who has that view of mathematics is unlikely to appreciate that research in mathematics is a thriving, worldwide activity, or to accept a suggestion that mathematics permeates, often to a considerable extent, most walks of present-day life and society. Nor are they likely to know which organization in the U.S. employs the greatest number of Ph.D.s in mathematics. (The answer is the National Security Agency. Exactly what the 30 or so new mathematics Ph.D.s who are hired there each year — the exact

number is an official secret — do to earn their paychecks is never made public, but it is generally assumed that the majority of them work on code breaking, to enable the agency to read encrypted messages that are intercepted by monitoring systems.)

In fact, the answer to the question "What is mathematics?" has changed several times during the course of history.

Up to 500 B.C. or thereabouts, mathematics was indeed the study of number. This was the period of Egyptian and Babylonian mathematics.[1] In those civilizations, mathematics consisted almost solely of arithmetic. It was largely utilitarian, and very much of a "cookbook" variety. ("Do such and such to a number and you will get the answer.")

The period from around 500 B.C. to 300 A.D. was the era of Greek mathematics. The mathematicians of ancient Greece were primarily concerned with geometry. Indeed, they regarded numbers in a geometric fashion, as measurements of length, and when they discovered that there were lengths to which their numbers did not correspond (the discovery of irrational lengths), their study of number largely came to a halt.[2] For the Greeks, with their emphasis on geometry, mathematics was the study of number and shape.

In fact, it was only with the Greeks that mathematics came into being as an area of study, and ceased being merely a collection of techniques for measuring, counting, and accounting. Greek interest in mathematics was not just utilitarian; they regarded mathematics as an intellectual pursuit having both aesthetic and religious elements. Around 500 B.C., Thales of Miletus (now part of Turkey) introduced the idea that the precisely stated assertions of mathematics could be logically proved by a formal argument. This innovation marked the birth of the theorem, now the bedrock of mathematics. For the Greeks, this approach culminated in the publication of Euclid's *Elements*, reputedly the most widely circulated book of all time after the Bible.

There was no major change in the overall nature of mathematics, and hardly any significant advances within the subject, until the middle of the 17th century, when Isaac Newton (in England) and Gottfried Leibniz (in Germany) independently invented calculus. In essence, calculus is the study of continuous motion and change. Previous mathematics had been largely restricted to the static issues of counting, measuring, and describing shape. With the introduction of techniques to handle motion and change, mathematicians were able to study the motion of the planets and of falling bodies on earth, the workings of machinery, the flow of liquids, the expansion of gases, physical forces such as magnetism and electricity, flight, the growth of plants and animals, the spread of epidemics, the fluctuation of profits, and so on.

[1] Other civilizations also developed mathematics, for example, the Chinese and the Japanese. But the mathematics of those cultures did not appear to have a direct influence on the development of modern Western mathematics, so in this book I will ignore them.

[2] There is an oft repeated story that the young Greek mathematician who made this discovery was taken out to sea and drowned, lest the awful news of what he had stumbled upon should leak out. As far as I know, there is no evidence whatsoever to support this fanciful tale. Pity. It's a great story.

After Newton and Leibniz, mathematics became the study of number, shape, motion, change, and space.

Most of the initial work involving calculus was directed toward the study of physics; indeed, many of the great mathematicians of the period are also regarded as physicists. But from about the middle of the 18th century there was an increasing interest in mathematics itself, not just its applications, as mathematicians sought to understand what lay behind the enormous power that calculus gave to humankind. Here the old Greek tradition of formal proof came back into ascendancy, as a large part of present-day pure mathematics was developed. By the end of the 19th century, mathematics had become the study of number, shape, motion, change, and space, and of the mathematical tools that are used in this study.

The explosion of mathematical activity that has taken place over the past 100 years or so has been dramatic. The growth has not just been a further development of previous mathematics; many quite new branches have sprung up. At the start of the 20th century, mathematics could reasonably be regarded as consisting of about 12 distinct subjects: arithmetic, geometry, calculus, and so on. Today, between 60 and 70 distinct categories would be a reasonable figure. Some subjects, such as algebra or topology, have split into various subfields; others, such as complexity theory or dynamical systems theory, are completely new areas of study.

Given this tremendous growth in mathematical activity, for a while it seemed as though the only simple answer to the question "What is mathematics?" was to say, somewhat fatuously, "It is what mathematicians do for a living." A particular study was classified as mathematics not so much because of what was studied but because of how it was studied — that is, the methodology used. It was only in the 1980s that a definition of mathematics emerged on which most mathematicians now agree: *mathematics is the science of patterns.* What the mathematician does is examine abstract patterns — numerical patterns, patterns of shape, patterns of motion, patterns of behavior, voting patterns in a population, patterns of repeating chance events, and so on. Those patterns can be either real or imagined, visual or mental, static or dynamic, qualitative or quantitative, purely utilitarian or of little more than recreational interest. They can arise from the world around us, from the depths of space and time, or from the inner workings of the human mind. Different kinds of patterns give rise to different branches of mathematics. For example:

- Arithmetic and number theory study the patterns of numbers and counting.

- Geometry studies the patterns of shape.

- Calculus allows us to handle patterns of motion.

- Logic studies patterns of reasoning.

- Probability theory deals with patterns of chance.

- Topology studies patterns of closeness and position.

and so forth.

1.2 Mathematical Notation

One aspect of modern mathematics that is obvious to even the casual observer is the use of abstract notations: algebraic expressions, complicated-looking formulas, and geometric diagrams. The mathematician's reliance on abstract notation is a reflection of the abstract nature of the patterns she studies.

Different aspects of reality require different forms of description. For example, the most appropriate way to study the lay of the land or to describe to someone how to find their way around a strange town is to draw a map. Text is far less appropriate. Analogously, line drawings in the form of blueprints are the appropriate way to specify the construction of a building. And musical notation is the most appropriate medium to convey music, apart from, perhaps, actually playing the piece.

In the case of various kinds of abstract, formal patterns and abstract structures, the most appropriate means of description and analysis is mathematics, using mathematical notations, concepts, and procedures. For instance, the symbolic notation of algebra is the most appropriate means of describing and analyzing general behavioral properties of addition and multiplication.

For example, the commutative law for addition could be written in English as:

When two numbers are added, their order is not important.

However, it is usually written in the symbolic form

$$m + n = n + m$$

Such is the complexity and the degree of abstraction of the majority of mathematical patterns, that to use anything other than symbolic notation would be prohibitively cumbersome. And so the development of mathematics has involved a steady increase in the use of abstract notations.

The first systematic use of a recognizably algebraic notation in mathematics seems to have been made by Diophantus, who lived in Alexandria some time around 250 A.D. His treatise *Arithmetica*, of which only six of the original thirteen volumes have been preserved, is generally regarded as the first algebra textbook. In particular, Diophantus used special symbols to denote the unknown in an equation and to denote powers of the unknown, and he employed symbols for subtraction and for equality.

These days, mathematics books tend to be awash with symbols, but mathematical notation no more is mathematics than musical notation is music. A page of sheet music *represents* a piece of music; the music itself is what you get when the notes on the page are sung or performed on a musical instrument. It is in its performance that the music comes alive and becomes part of our experience; the music exists not on the printed page but in our minds. The same is true for mathematics; the symbols on a page are just a *representation* of the mathematics. When read by a competent performer (in this case, someone trained in mathematics), the symbols on the printed page come alive — the mathematics lives and breathes in the mind of the reader like some abstract symphony.

Given the strong similarity between mathematics and music, both of which have their own highly abstract notations and are governed by their own structural rules, it is hardly surprising that many (perhaps most) mathematicians also have some musical talent. Although some commentators make more of this connection in terms of mental ability than I believe is warranted, it is true that for most of the 2,500 years of Western civilization, starting with the ancient Greeks, mathematics and music were regarded as two sides of the same coin. It was only with the rise of the scientific method in the 17th century that the two started to go their separate ways.

For all the historical connections, however, there was, until recently, one very obvious difference between mathematics and music. Although only someone well trained in music can read a musical score and hear the music in her head, if that same piece of music is performed by a competent musician, anyone with a sense of hearing can appreciate the result. It requires no musical training to experience and enjoy music when it is performed.

For most of its history, however, the only way to appreciate mathematics was to learn how to "sight-read" the symbols. Although the structures and patterns of mathematics reflect the structure of, and resonate in, the human mind every bit as much as do the structures and patterns of music, human beings have developed no mathematical equivalent to a pair of ears. Mathematics can only be "seen" with the "eyes of the mind". It is as if we had no sense of hearing, so that only someone able to sight-read musical notation would be able to appreciate the patterns and harmonies of music.

In recent years, however, the development of computer and video technologies has to some extent made mathematics accessible to the untrained spectator. In the hands of a skilled user, the computer can be used to "perform" mathematics, and the result can be displayed in a visual form on the screen for all to see. Although only a relatively small part of mathematics lends itself to such visual "performance", it is now possible to convey to the layperson at least something of the beauty and the harmony that the mathematician "sees" and experiences when she does mathematics.

Sometimes, the use of computer graphics can be of significant use to the mathematician as well as providing the layperson with a glimpse of the inner world of mathematics. For instance, the study of complex dynamical systems was begun in the 1920s by the French mathematicians Pierre Fatou and Gaston Julia, but it was not until the late 1970s and early 1980s that the rapidly developing technology of computer graphics enabled Benoit Mandelbrot and other mathematicians to see some of the structures Fatou and Julia had been working with. The strikingly beautiful pictures that emerged from this study have since become something of an art form in their own right.

Without its algebraic symbols, large parts of mathematics simply would not exist. Indeed, the issue is a deep one having to do with human cognitive abilities. The recognition of abstract concepts and the development of an appropriate language are really two sides of the same coin.

The use of a symbol such as a letter, a word, or a picture to denote an abstract entity goes hand in hand with the recognition of that entity as an entity. The use of the numeral "7" to denote the number 7 requires that the number 7 be recognized as

an entity; the use of the letter m to denote an arbitrary whole number requires that the concept of a whole number be recognized. Having the symbol makes it possible to think about and manipulate the concept.

This linguistic aspect of mathematics is often overlooked. Indeed, one often hears the complaint that mathematics would be much easier if it weren't for all that abstract notation, which is rather like saying that Shakespeare would be much easier to understand if it were written in simpler language.

Sadly, the level of abstraction in mathematics, and the consequent need for notations that can cope with that abstraction, means that many, perhaps most, parts of mathematics will remain forever hidden from the non-mathematician; and even the more accessible parts may be at best dimly perceived, with much of their inner beauty locked away from view.

Inner beauty? What do I mean by that? In what sense is mathematics beautiful?

Mathematical beauty is hard to convey to an outsider. Apart from computer graphical representations of a few mathematical objects, where there is a visual beauty plain for all to see, mathematical beauty is highly abstract — a beauty of abstract form, structure, and logic.

In his 1940 book *A Mathematician's Apology*, the accomplished English mathematician G. H. Hardy wrote:

> The mathematician's patterns, like the painter's or the poet's, must be beautiful, the ideas, like the colors or the words, must fit together in a harmonious way. Beauty is the first test; there is no permanent place in the world for ugly mathematics. ...It may be very hard to define mathematical beauty, but that is just as true of beauty of any kind — we may not know quite what we mean by a beautiful poem, but that does not prevent us from recognizing one when we read it.

The beauty to which Hardy was referring is a highly abstract, inner beauty, a beauty of abstract form and logical structure, a beauty that can be observed, and appreciated, only by those sufficiently well trained in the discipline. It is a beauty "cold and austere", according to Bertrand Russell, the famous English mathematician and philosopher, who wrote in his 1918 book *Mysticism and Logic*:

> Mathematics, rightly viewed, possesses not only truth, but supreme beauty — a beauty cold and austere, like that of sculpture, without appeal to any part of our weaker nature, without the gorgeous trappings of painting or music, yet sublimely pure, and capable of a stern perfection such as only the greatest art can show.

Mathematics, the science of patterns, is a way of looking at the world, both the physical, biological, and sociological world we inhabit and the inner world of our minds and thoughts. Mathematics' greatest success has undoubtedly been in the physical domain, where the subject is rightly referred to as both the queen and the servant of the (natural) sciences. Yet, as an entirely human creation, the study of mathematics is ultimately a study of humanity itself. For none of the entities that form the substrate of mathematics exists in the physical world; the numbers, the points, the lines and planes, the surfaces, the geometric figures, the functions, and so forth

are pure abstractions that exist only in humanity's collective mind. The absolute certainty of a mathematical proof and the indefinitely enduring nature of mathematical truth are reflections of the deep and fundamental status of the mathematician's patterns in both the human mind and the physical world.

In an age when the study of the heavens dominated scientific thought, Galileo said,

> The great book of nature can be read only by those who know the language in which it was written. And this language is mathematics.

Striking a similar note in a much later era, when the study of the inner workings of the atom had occupied the minds of many scientists for a generation, the Cambridge physicist John Polkinhorne wrote, in 1986,

> Mathematics is the abstract key which turns the lock of the physical universe.

In today's age, dominated by information, communication, and computation, mathematics is finding new locks to turn. As the science of abstract patterns, there is scarcely any aspect of our lives that is not affected, to a greater or lesser extent, by mathematics; for abstract patterns are the very essence of thought, of communication, of computation, of society, and of life itself.

1.3 Making the Invisible Visible

We have answered the question "What is mathematics?" with the catch phrase "Mathematics is the science of patterns." The second fundamental question about mathematics that we posed at the start of the chapter — "What does it do for us?" — can also be answered with a catchy phrase: *Mathematics makes the invisible visible.*

Let me give you some examples of what I mean by this answer.

Without mathematics, there is no way you can understand what keeps a jumbo jet in the air. As we all know, large metal objects don't stay above the ground without something to support them. But when you look at a jet aircraft flying overhead, you can't see anything holding it up. It takes mathematics to "see" what keeps an airplane aloft. In this case, one way to "see" the invisible is an equation discovered by the mathematician Daniel Bernoulli early in the 18th century.

While I'm on the subject of flying, what is it that causes objects other than aircraft to fall to the ground when we release them? "Gravity," you answer. But that's just giving it a name. It doesn't help us to understand it. It's still invisible. We might as well call it magic. To understand it, you have to "see" it. That's exactly what Newton did with his equations of motion and mechanics in the 17th century. Newton's mathematics enabled us to "see" the invisible forces that keep the earth rotating around the sun and cause an apple to fall from the tree onto the ground.

Both Bernoulli's equation and Newton's equations use calculus. Calculus works by making visible the infinitesimally small. That's another example of making the invisible visible.

Here's another: Two thousand years before we could send spacecraft into outer space to provide us with pictures of our planet, the Greek mathematician Eratosthenes used mathematics to show that the Earth was round. Indeed, he calculated its diameter, and hence its curvature, with 99% accuracy.

Using mathematics, we have been able to look backward and see into the distant past, making visible the otherwise invisible moments when the universe was first created in what we call the Big Bang.

For a more commonplace example, how do you "see" what makes pictures and sound of a football game miraculously appear on a television screen on the other side of town? One answer is that the pictures and sound are transmitted by radio waves — a special case of what we call electromagnetic radiation. But as with gravity, that just gives the phenomenon a name, it doesn't help us to "see" it. In order to "see" radio waves, you have to use mathematics. Maxwell's equations, discovered in the 19th century, make visible to us the otherwise invisible radio waves.

Here are some human patterns:

- Aristotle used mathematics to try to "see" the invisible patterns of sound that we recognize as music.

- He also used mathematics to try to describe the invisible structure of a dramatic performance.

- In the 1950s, the linguist Noam Chomsky used mathematics to "see" and describe the invisible, abstract patterns of words that we recognize as a grammatical sentence. He thereby turned linguistics from a fairly obscure branch of anthropology into a thriving mathematical science.

Finally, using mathematics, we are able to look into the future:

- Probability theory and mathematical statistics let us predict the outcomes of elections, often with remarkable accuracy.

- We use calculus to predict tomorrow's weather.

- Market analysts use various mathematical theories to try to predict the future behavior of the stock market.

- Insurance companies uses statistics and probability theory to predict the likelihood of an accident during the coming year, and set their premiums accordingly.

When it comes to looking into the future, mathematics allows us to make visible another invisible — the invisible of the not yet happened. In that case, our mathematical vision is not perfect. Our predictions are sometimes wrong. But without mathematics, we cannot even see poorly.

Today we live in a technological society, built on mathematics. There are increasingly fewer places on the face of the earth where, when we look around us toward the horizon, we do not see products of our technology — and hence of our

mathematics: tall buildings, bridges, power lines, telephone cables, cars on roads, aircraft in the sky, and so forth. Where communicating with one another once required physical proximity, today much of our communication is mediated by mathematics, transmitted in digitized form along wires or optical fibers, or through the ether. Computers — machines that perform mathematics — are not only on our desktops, they are in everything from cell phones to microwave ovens, from interactive games to automobiles, and from children's toys to heart pacemakers. Mathematics — in the form of statistics — is used to decide what food we will eat, what products we will buy, what television programs we will be able to see, and which politicians we will be able to vote for. Just as society burned fossil fuels to drive the engines of the industrial age, in today's information age, the principal fuel we burn is mathematics.

And yet, as the role of mathematics in our lives has grown more and more significant, it has become more and more hidden from view, forming an invisible universe that supports much of our lives. Just as our every action is governed by the invisible forces of nature (such as gravity), so too we now live in the invisible universe created by mathematics, subject to invisible mathematical laws.

1.4 This Is Where You Come In

That then, is the big picture of mathematics. Hopefully now you have a good overall sense of what the subject is and why it is important, even if you did not previously. But what does it mean to *do* mathematics? What is it that you, the beginning student of post-calculus math, must learn to do? Here too, the answer we give today is not the one that would have been given in the past.

Up to about 150 years ago, although mathematicians had long ago expanded the realm of objects they studied beyond numbers and algebraic symbols for numbers, they still regarded mathematics as primarily about *calculation*. That is, proficiency in mathematics depended above all at being able to carry out calculations or manipulate symbolic expressions to solve problems. By and large, high school mathematics is still very much based on that earlier tradition.

In the middle of the 19th century, however, a revolution took place. One of its epicenters was the small university town of Göttingen in Germany, where the local revolutionary leaders were the mathematicians Lejeune Dirichlet, Richard Dedekind, and Bernhard Riemann. In their new conception of the subject, the primary focus was not performing a calculation or computing an answer, but formulating and understanding abstract concepts and relationships. This was a shift in emphasis from *doing* to *understanding*. Within a generation, this revolution would completely change the way pure mathematicians thought of their subject. Nevertheless, it was an extremely quiet revolution that was recognized as such only when it was all over. It is not even clear that the leaders knew they were spearheading a major change.

For the Göttingen revolutionaries, mathematics was about "thinking in concepts" (*Denken in Begriffen*). Mathematical objects were no longer thought of as given primarily by formulas, but rather as carriers of conceptual properties. Proving was

no longer a matter of transforming terms in accordance with rules, but a process of logical deduction from concepts.

This new approach to mathematics now permeates all university mathematics instruction beyond calculus. And by and large, it is the post-Göttingen approach to mathematics that this little book is all about. Among the new concepts that the revolution embraced are many with which today's university mathematics student must come to grips. You will learn about some of them in the pages that follow.

One such post-Göttingen concept is that of a *function*. Prior to the 19th century, mathematicians were used to the fact that a formula such as $y = x^2 + 3x - 5$ specifies a rule that produces a new number (y) from any given number (x). Then along came the Göttingen revolutionary Dirichlet who said, forget the formula and concentrate on what the function *does* in terms of input–output behavior. A *function*, according to Dirichlet, is any rule that produces new numbers from old. The rule does not have to be specified by an algebraic formula. In fact, there's no reason to restrict your attention to numbers. A function can be any rule that takes objects of one kind and produces new objects from them.

Mathematicians began to study the properties of abstract functions, specified not by some formula but by their behavior. For example, does the function have the property that when you present it with different starting values it always produces different answers? (This property is called *injectivity*. We'll meet it again later.)

This approach was particularly fruitful in the development of real analysis, where mathematicians studied the properties of continuity and differentiability of functions as abstract concepts in their own right. In France, Augustin Cauchy developed his famous (some might say infamous) epsilon-delta definitions of continuity and differentiability — the *epsilontics* that to this day cost each new generation of post-calculus mathematics students so much effort to master. Cauchy's contributions, in particular, indicated a new willingness of mathematicians to come to grips with the concept of infinity. Riemann spoke of their having reached "a turning point in the conception of the infinite".

In the 1850s, Riemann defined a complex function *by its property of differentiability*, rather than a formula, which he regarded as secondary. Karl Friedrich Gauss' residue classes (which you are likely to meet in an algebra course) were a forerunner of the approach — now standard — whereby a mathematical structure is defined as a set endowed with certain operations, whose behaviors are specified by axioms. Taking his lead from Gauss, Dedekind examined the new concepts of ring, field, and ideal — each of which was defined as a collection of objects endowed with certain operations. (Again, these are concepts you are likely to encounter soon in your post-calculus mathematics education.)

Like most revolutions, the Göttingen one had its origins long before the main protagonists came on the scene. The Greeks had certainly shown an interest in mathematics as a conceptual endeavor, not just calculation, and in the 17th century, Gottfried Leibniz thought deeply about both approaches. But for the most part, until the Göttingen revolution, mathematics was viewed primarily as a collection of procedures for solving problems. To today's mathematicians, however, brought up entirely with the post-Göttingen conception of mathematics, what in the 19th century was a

revolution is simply taken to be what mathematics is. The revolution may have been quiet, and to a large extent forgotten, but it was complete and far reaching. And it sets the scene for this book, the main aim of which is to provide you with the basic tools you will need to enter this new world of modern mathematics.

Although the Göttingen view of mathematics now dominates the field at the post-calculus university level, it has not had much infuence on high school mathematics — which is why you need this little book to help you make the transition. There was one attempt to introduce the new approach into school classrooms, but it went terribly wrong and soon had to be abandoned. This was the so-called "New Math" movement of the 1960s. What went wrong was that by the time the Göttingen message had made its way from the mathematics departments of the leading universities into the schools, it was badly garbled. To mathematicians before and after 1850, both calculation and understanding had always been important. The 1850 revolution merely shifted the *emphasis* as to which of the two the subject was really about and which was the supporting skill. Unfortunately, the message that reached the nation's school teachers in the 1960s was often, "Forget calculation skill, just concentrate on concepts." This ludicrous and ultimately disastrous strategy led the satirist Tom Lehrer to quip, in his song *New Math*, "It's the method that's important, never mind if you don't get the right answer." (Lehrer, by the way, is a mathematician, so he knew what the initiators of the change had intended.) After a few sorry years, the "New Math" (which was already over 100 years old) was dropped from the syllabus. Such is the nature of educational policy making in free societies that it is unlikely that such a change could ever be made in the foreseeable future, even if it were done properly the second time around. It's also not clear (at least to me) that such a change would be altogether desirable. There are educational arguments (which in the absense of hard evidence either way are hotly debated) that say the human mind has to achieve a certain level of mastery of computation with abstract mathematical entities before it is able to reason about their properties.

1.5 The Stuff of Modern Mathematics

In modern, pure mathematics we are primarily concerned with *statements* about *mathematical objects*.

Mathematical objects are things such as integers, real numbers, sets, functions, etc. Examples of mathematical statements are:

(1) There are infinitely many prime numbers.

(2) For every real number a, the equation $x^2 + a = 0$ has a real root.

(3) $\sqrt{2}$ is irrational.

(4) If $p(n)$ denotes the number of primes less than or equal to the natural number n, then as n becomes very large, $p(n)$ approaches $n / \log_e n$.

Not only are we interested in statements of the above kind, we are, above all, interested in knowing which statements are true and which are false. (The truth or falsity in each case is demonstrated not by measurement or experiment, as in most other sciences, but by a *proof*, of which more in due course). For instance, in the above examples, (1), (3), and (4) are true whereas (2) is false.

The truth of (1) is easily proved. We show that if we list the primes in increasing order as

$$p_1, p_2, p_3, \ldots, p_n, \ldots$$

then the list must continue forever. We all know what the first few members of the sequence are: $p_1 = 2$, $p_2 = 3$, $p_3 = 5$, $p_4 = 7$, $p_5 = 11, \ldots$ Consider the list up to stage n:

$$p_1, p_2, p_3, \ldots, p_n$$

Let

$$p = (p_1.p_2.p_3.\ldots.p_n) + 1$$

If p is not a prime, there must be a prime $q < p$ such that q divides p. But none of p_1, \ldots, p_n divides p, for the division of p by any one of these leaves a remainder of 1. So, either p must itself be prime, or else there is a prime $q < p$ that exceeds p_n. Either way we see that there is a prime greater than p_n. Since this argument does not depend in any way upon the size of n, it follows that there are infinitely many primes.

Example (2) can easily be proved to be false. Since the square of no real number is negative, the equation $x^2 + 1 = 0$ does not have a real root.

We give a proof of (3) later. The only known proofs of example (4) are extremely complicated — far too complicated to be included in an introductory text such as this.

Clearly, before we can prove whether certain statements are true or false, we must be able to understand precisely what a statement says. Above all, mathematics is a very *precise* subject, where exactness of expression is required. This already creates a difficulty, for in real life our use of language is rarely precise. Now, systematically to make the entire English language precise (by defining *exactly* what each word is to mean) would be an impossible task. It would also be unnecessary. It turns out that by deciding exactly what we mean by a few simple words, we can obtain enough precision to enable us to make all (or at least most) of the mathematical statements we need. The point is that in mathematics, we do not use all of the English language. Indeed, when we restrict our attention to mathematical *statements* themselves (as opposed to our attempts to explain them), we need only a very small part of the English language.

Chapter 2

Math Speak

2.1 The Language of Mathematics: Part 1

Language — ordinary human language such as English — is a remarkably efficient way of communicating. To the best of our knowledge, our ancestors acquired the ability to use language about 100,000 years ago, which means it is a very recent ability in evolutionary terms. Much of the efficiency of language comes from a feature that linguists refer to as *indexicality*. This means the way the meaning of things we say or write depends upon both the context in which we say or write them and the context in which what we say or write is heard or read. For example, if I say "I am tired" it means that the mathematician and author Keith Devlin is tired. But if you say the very same sentence, it means something different, namely that *you* are tired. For another example, take the word "small". Its meaning in the phrase "small rodent" is different from its meaning in the phrase "small elephant". Most people would agree that a small elephant is definitely not a small animal. "Summer months" means June, July, and August in the northern hemisphere and December, January, and February in the southern hemisphere. Indexicality — the dependency of language on context to determine the meaning — is what makes it possible for a relatively small collection of words to enable us to talk about a much larger range of topics.

For examples of how context affects meaning in more complex situations, consider the following examples of ambiguous sentences, where the meaning depends on the context:

- The man saw the woman with a telescope.

- The two sisters were reunited after 10 years in the checkout line at Safeway.

- Two Catholic prostitutes appealed to the Pope.

- A large hole appeared in Main Street. The police are looking into it.

- The mayor said that bus passengers should be belted.

Unfortunately, the indexicality of language, while of great use in everyday communication — and, as the above examples show, occasionally a source of some amusement — can be problematical when it comes to talking or writing

about mathematics. Mathematical statements are supposed to have a unique meaning, independent of context.

Now, much of the ambiguity of language disappears once we use language to talk about mathematics. In fact, most does. Most, but not all. There are a few key words that are ambiguous even when used to express mathematical facts. They are the words used to modify or combine statements to form more complex assertions — words such as "and", "or", and "not". What mathematicians have done — and much of this was done by the ancient Greeks over 2,000 years ago — is to specify at the outset exactly which of the possible meanings is intended whenever the word is used in mathematics.

Unfortunately for the beginner, the mathematical usage of some of these words is not quite the same as the everyday, non-mathematical usage. But there is no major problem with the first word we consider:

and

We need to be able to assert that two events hold simultaneously. For instance, we may wish to say that π is greater than 3 *and* less than 3.2. So the word *and* is indispensable. Sometimes, in order to achieve a completely symbolic expression, we introduce an abbreviation for *and*. The most common ones are

$$\wedge \quad , \quad \&$$

Thus, the expression

$$(\pi > 3) \wedge (\pi < 3.2)$$

says:

$$\pi \text{ is greater than 3 } \textit{and } \pi \text{ is less than 3.2}$$

In other words, π lies between 3 and 3.2.

There is no possible source of confusion when we use the word *and*. If ϕ and ψ are any two mathematical statements,

$$\phi \wedge \psi$$

is the joint assertion (which may or may not be a valid assertion) of *both ϕ and ψ*.

One thing to notice is that in mathematics, *and* is independent of order: $\phi \wedge \psi$ means the same as $\psi \wedge \phi$. This is not always true when we use *and* in everyday life. For example,

John took the free kick and the ball went into the net

does not mean the same as

The ball went into the net and John took the free kick.

Ambiguity arises in the case of the next word we look at:

or

We wish to be able to assert that event A occurs *or* event B occurs. For instance, we might want to say

$$a > 0 \quad or \quad \text{the equation } x^2 + a = 0 \text{ has a real root}$$

or perhaps we want to say

$$ab = 0 \quad if \quad a = 0 \quad or \quad b = 0$$

The use of *or* is different in these two examples. In the first assertion there is no possibility of both eventualities occurring at once. Moreover, the meaning is unchanged if we put the word *either* at the beginning of the sentence. In the second case, it is quite possible for *both a* and *b* to be zero.[1]

But mathematics has no place for potential ambiguity in the meaning of such a commonly used word as *or*, so we must choose one or the other meaning. It turns out to be more convenient to opt for the inclusive use. Accordingly, whenever we use the word *or* in mathematics we always mean the *inclusive or*. If ϕ and ψ are mathematical statements, ϕ *or* ψ asserts that at least one of ϕ, ψ is valid. We often use the symbol

$$\vee$$

to denote (*inclusive*) *or*. Thus

$$\phi \vee \psi$$

means that *at least one of* ϕ, ψ is valid.

For instance, the following (rather silly) statement is true:

$$(3 < 5) \vee (1 = 0)$$

(For all that this is a silly example, you should pause for a moment and make sure you understand why this statement is not only mathematically *meaningful* but actually true.)

Our next word arises by virtue of our need to negate statements, to say that a particular statement is false. So we need to be sure of our use of the word

not

If ψ is any statement,

$$not - \psi$$

claims that ψ is false. Thus, if ψ is a true statement, *not*–ψ is a false statement, and if ψ is a false statement, *not*–ψ is a true statement. Nowadays, the symbol most commonly used for *not* is

$$\neg$$

but older texts sometimes use the symbol \sim.

[1] Even if we were to alter our second assertion by inserting the word *either*, we would still read it as allowing for both possibilities to occur at once, since the use of the word *either* only serves to strengthen the exclusivity in an assertion *where it is already clear* that there is no possibility of both.

Although our usage of the word *not* accords with most common usage, negation is sometimes used very loosely in everyday speech. For instance, there is no confusion about the meaning of the statement

$$\neg(\pi < 3)$$

This clearly means

$$\pi \geq 3$$

which, incidentally, is the same as $(\pi = 3) \vee (\pi > 3)$. But consider the statement

All domestic cars are badly made.

What is the negation of this statement? For instance, is it any one of the following?

 (a) All domestic cars are well made.

 (b) All domestic cars are not badly made.

 (c) At least one domestic car is well made.

 (d) At least one domestic car is not badly made.

A common mistake is for the beginner to choose (a). But this is easily seen to be wrong. Our original statement is surely false. Hence the *negation* of this statement will be true. But (a) is certainly not true! Neither is (b) true. So realistic considerations lead us to conclude that if the correct answer is to be found in the above list, then it has to be either (c) or (d). (We shall later see how we can eliminate (a) and (b) by a formal mathematical argument.)

 In point of fact, both (c) and (d) can be said to represent the negation of our original statement. (Any well made domestic car testifies the *truth* of both (c) and (d).) Which do you think most closely corresponds to the negation of our original statement?

 We shall return to this example later, but before we leave it for now, let us note that the original statement is only concerned with domestic cars. Hence its negation will only deal with domestic cars. So the negation will not involve any reference to *foreign* cars. For instance, the statement

All foreign cars are well made

cannot be the negation of our original statement. Indeed, knowing whether our original statement is true or false in no way helps us to decide the truth or falsity of the above statement. To be sure, "foreign" is the negation of "domestic", but we are trying to negate the assertion as a whole, not some adjective occurring in it.

 So far we have considered three basic words and made their meaning precise: *and* (symbols: \wedge, &), *or* (symbol: \vee), and *not* (symbol: \neg). We refer to these operations as *conjunction, disjunction,* and *negation,* respectively. Thus $\phi \wedge \psi$ is the *conjunction* of the statements ϕ and ψ, $\phi \vee \psi$ is their *disjunction,* and $\neg\phi$ is the *negation* of ϕ.

The fourth notion we investigate is the one that causes by far the most initial confusion: *implication*. In mathematics we frequently use an expression of the form

$$(*) \quad \phi \; implies \; \psi$$

Indeed implication provides the means by which we prove statements, starting from initial observations or axioms. The question is, what is the meaning of an assertion of the form $(*)$?

It would not be unreasonable to assume it means the following:

If ϕ *is true, then* ψ *has to be true as well*

But suppose that for ϕ we take the true assertion "$\sqrt{2}$ is irrational" and for ψ we take the true assertion "$0 < 1$". Then is the expression $(*)$ true? In other words, does the irrationality of $\sqrt{2}$ imply that 0 is less than 1? Of course it does not. There is no real connection between the two statements ϕ and ψ in this case.

Even more dramatic, what about the implication

$$(\text{Julius Caesar is dead}) \; implies \; (1 + 1 = 2)$$

Again, what can be said about the truth of $(*)$ if ϕ is false? Does the expression have any meaning in such a case?

It all seems very confusing. The problem is that the concept of implication involves not just truth (as was the case with the words *and*, *or*, *not* we have considered already), but also *causation*. When we say that "ϕ implies ψ", we mean that ϕ somehow causes or brings about ψ. This certainly has the consequence that the truth of ψ follows from the truth of ϕ, but *truth alone does not fully capture what is going on*.

For our present purposes, namely the introduction of some precision into our use of language in mathematics, the complex issue of causation is not one that we wish to consider. So what we shall do is pull apart the notion of implication into two parts, the truth part and the causation part, and discard the latter. The part that we are left with, namely the 'truth part' of implication, is generally known as the *conditional*, or sometimes the *material conditional*. Thus we have the relationship:

$$\text{implication} = \text{conditional} + \text{causation}$$

We shall use the notation

$$\phi \Rightarrow \psi$$

to denote the truth-value, *conditional* part of the implication $(*)$. (Modern logic texts generally use a single arrow, \rightarrow, instead of \Rightarrow, but to avoid confusion with the notation for functions we shall introduce later, we shall adopt the more old-fashioned, double-arrow notation here.)

Any expression of the form

$$\phi \Rightarrow \psi$$

will be referred to as *a conditional expression*, or simply *a conditional*. We refer to ϕ as the *antecedent* of the conditional and ψ as the *consequent*. The truth or falsity of a conditional will be *defined* entirely in terms of the truth or falsity of the antecedent and the consequent. That is to say, whether or not the conditional expression $\phi \Rightarrow \psi$ is true will depend entirely upon the truth or falsity of ϕ and ψ, taking no account of whether or not there is any meaningful connection between ϕ and ψ.

As you might expect, such a definition, ignoring as it does a highly significant aspect of the notion of implication, is likely to have consequences that are at the very least counter-intuitive, and possibly even absurd. However (and this is why the approach we adopt turns out to be a useful one), in all cases where there *is* a meaningful and genuine implication ϕ *implies* ψ, the conditional $\phi \Rightarrow \psi$ does accord with that implication.

That is to say, in defining the conditional the way we do, we do not end up with a notion that *contradicts* the notion of (genuine) implication. Rather we obtain a notion that *extends* (genuine) implication to cover those cases where a claim of implication is meaningless (namely, cases where the antecedent is false or there is no real connection between the antecedent and the consequent). In all cases where ϕ actually does imply ψ, the truth value (i.e., true or false) of the conditional will be the same as the truth value of the actual implication.

It is the fact that the conditional always has a well-defined truth value that makes this notion important in mathematics, since in mathematics we cannot afford to have statements with undefined truth values floating around.

Now, ignoring all questions of causality, the truth-value aspect of implication seems straightforward enough when the antecedent is true. If ϕ is true and "ϕ *implies* ψ" is a valid implication, then ψ will be true. Thus the conditional $\phi \Rightarrow \psi$ should be true if both the antecedent and consequent are true. But what about the case where the antecedent is false?

To deal with this case, we consider not the notion of implication, but its negation. We extract the causation-free, truth-value part of the statement "ϕ does not imply ψ", which we shall write as

$$\phi \not\Rightarrow \psi$$

Leaving aside all question of whether there is a meaningful causal relation between ϕ and ψ, and concentrating solely on truth values, how can we be sure that $\phi \not\Rightarrow \psi$ is a valid statement? More precisely, how should the truth or falsity of the assertion $\phi \not\Rightarrow \psi$ depend upon the truth or falsity of ϕ and ψ?

Well, ϕ will *not* imply ψ if it is the case that *although* ϕ is true, ψ is *nevertheless* false.

Please read that last sentence again.

Now once more.

Okay, now we are ready to continue.[2]

We therefore define $\phi \not\Rightarrow \psi$ to be true precisely in case ϕ is true and ψ is false.

[2]On second thought, maybe you should read it a fourth time just to be sure.

Having defined the truth or falsity of $\phi \not\Rightarrow \psi$, we obtain that of $\phi \Rightarrow \psi$ by just taking the negation. The conditional $\phi \Rightarrow \psi$ will be true exactly when $\phi \not\Rightarrow \psi$ is false. Examination of this definition leads to the following conclusions: $\phi \Rightarrow \psi$ will be true whenever one of the following holds:

(1) ϕ and ψ are both true.

(2) ϕ and ψ are both false.

(3) ϕ is false and ψ is true.

The points to note about this are:

(a) We are *defining* a notion that only captures part of what "implies" means.

(b) To avoid difficulties, we base our definition solely on the notion of truth and falsity.

(c) Our definition agrees with our intuition concerning implication in all meaningful cases.

Closely related to implication is the notion of *equivalence*. Two statements ϕ and ψ are said to be *(logically) equivalent* if each implies the other. The analogous, formal notion defined in terms of the conditional is known as the *biconditional*. We shall write the biconditional as

$$\phi \Leftrightarrow \psi$$

(Modern logic texts use $\phi \leftrightarrow \psi$). The biconditional is defined to be an abbreviation for the conjunction

$$(\phi \Rightarrow \psi) \wedge (\psi \Rightarrow \phi)$$

Looking back at the definition of the conditional, this means that the biconditional $\phi \Leftrightarrow \psi$ will be true if ϕ and ψ are both true or both false; otherwise (i.e., if exactly one of ϕ, ψ is true and the other false) the biconditional will be false. The three remarks (a), (b), (c) above apply to the connection between the biconditional and equivalence just as to the relation between the conditional and implication.

Exercises 2.1

(1) Simplify the following statements as much as you can:

(a) $(\pi > 0) \wedge (\pi < 10)$

(b) $(3 < 4) \wedge (3 < 6)$

 (c) $(e < 4) \wedge (e^2 < 9)$

 (d) $(\pi > 0) \vee (\pi > 1)$

 (e) $(\pi < 0) \vee (\pi > 0)$

(2) Which of the following are true and which are false?

 (a) $(\pi^2 > 2) \Rightarrow (\pi > 1.4)$

 (b) $(\pi^2 < 0) \Rightarrow (\pi = 3)$

 (c) $(\pi^2 > 0) \Rightarrow (1 + 2 = 4)$

 (d) $(\pi < \pi^2) \Rightarrow (\pi = 5)$

 (e) $(e^2 \geq 0) \Rightarrow (e < 0)$

 (f) $\neg[(5 \text{ is an integer}) \Rightarrow (5^2 \geq 1)]$

 (g) (the area of a circle of radius 1 is π) \Rightarrow (3 is a prime)

 (h) If squares have three sides, then triangles have four sides.

 (i) If elephants can climb trees, then 3 is an irrational number.

 (j) If Euclid's birthday was July 4, then rectangles have four sides.

2.2 Properties of the Language

So far we have taken five common words, *and, or, not, implies, equivalent*, and have attempted to make their meanings precise. We obtained the precision we wanted by basing our definitions on truth and falsity only, not on any causal relations or other connections the words might also convey. In the case of the first three, *and, or, not*, the precise definitions we gave accord reasonably well with common usage of the words outside of mathematics. In the case of the last two on the list, *implies* and *equivalent*, we defined precise, formal counterparts to implication and equivalence, called the conditional and the biconditional, respectively, that have some properties in common with the everyday meanings of *implies* and *equivalent* but in other ways are quite different.

We also introduced mathematical symbols to denote our formally defined concepts: \wedge for *and*, \vee for *or*, \neg for *not*, \Rightarrow for the conditional, and \Leftrightarrow for the biconditional. Whenever unambiguous precision is necessary, we can use these symbols with absolute conviction: there should be no possible cause for confusion (except due to misunderstanding of the concepts, a problem that can only be overcome by experience on the part of you, the student).

For an obvious reason, \wedge, \vee, \neg, \Rightarrow, and \Leftrightarrow are called *logical connectives*.

There are two further concepts that we take from everyday speech, but we postpone looking at those until later. In the meantime we consider the language developed so far. That is, we investigate the consequences of our formal definitions, and also introduce some associated terminology that is often used.

Negation

We consider first the behavior of negation. Our first observation is very simple. We note that $\neg(\neg\phi)$ is equivalent to (i.e., means the same as) ϕ.

This property of negation is not necessarily true for negation in everyday life. For example, you might find yourself saying "I was not displeased with the movie." In terms of negation, this has the form $\neg(\neg\text{ PLEASED})$, but your statement clearly does not mean that you were pleased with the movie. Indeed, it very definitely means something considerably less positive.

Figuring out the effect of negation on conjunctions and disjunctions takes a bit more effort. For instance, what is the meaning of an expression such as

$$\neg(\phi \wedge \psi)$$

where ϕ and ψ are given mathematical statements?

Well, $\phi \wedge \psi$ means "both ϕ and ψ are true", so $\neg(\phi \wedge \psi)$ means it is not the case that both ϕ and ψ are true. Now, if they are not both true, then at least one of them must be false. But to say at least one of ϕ and ψ is false is clearly the same as saying that at least one of $\neg\phi$ and $\neg\psi$ is true (by our definition of negation). By our meaning for *or*, this can be expressed as $(\neg\phi) \vee (\neg\psi)$. Thus

$$\neg(\phi \wedge \psi) \quad \text{and} \quad (\neg\phi) \vee (\neg\psi)$$

are equivalent (i.e., mean the same).

Likewise it can be shown (Exercise: carry out the required logical argument) that

$$\neg(\phi \vee \psi) \quad \text{and} \quad (\neg\phi) \wedge (\neg\psi)$$

are equivalent.

Thus, negation has the effect that it changes \vee into \wedge and changes \wedge into \vee. Knowing this now, you should reread the above proof that $\neg(\phi \wedge \psi)$ and $(\neg\phi) \vee (\neg\psi)$ are the same.

The effect of negation on conditionals is determined by the manner in which we *defined* the conditional in terms of negatives. By our definition

$$\neg(\phi \Rightarrow \psi)$$

is the same as

$$\phi \wedge (\neg\psi)$$

and (hence)

$$\phi \Rightarrow \psi$$

is the same as

$$(\neg\phi) \vee \psi$$

Of course, once we have made the meanings of various words precise, and introduced symbolic abbreviations, we are free to alter our notation if it seems helpful. For instance, we usually write

$$\phi \nRightarrow \psi$$

instead of

$$\neg(\phi \Rightarrow \psi)$$

Likewise, we use the more familiar

$$x \neq y$$

instead of

$$\neg(x = y)$$

Another example: in dealing with real numbers, we usually write, say,

$$a < x \leq b$$

instead of

$$(a < x) \wedge (x \leq b)$$

But we would probably write

$$\neg(a < x \leq b)$$

instead of

$$a \not< x \not\leq b$$

as the latter would appear to be rather unclear as to its meaning. (We could make this precise, but it still seems rather inelegant, and mathematicians don't do it.)

Implication

There is some terminology associated with implication (i.e., real implication, not just the conditional) that should be mastered straightaway as it pervades all mathematical discussion.

In an implication

$$\phi \text{ implies } \psi$$

we call ϕ the *antecedent* and ψ the *consequent*.

The following all mean the same:

(1) ϕ implies ψ

(2) if ϕ then ψ

(3) ϕ is sufficient for ψ

(4) ϕ only if ψ

(5) ψ if ϕ

(6) ψ whenever ϕ

(7) ψ is necessary for ϕ

The first four all mention ϕ before ψ, and of these the first three seem obvious enough. But caution is required in the case of (4). Notice the contrast between (4) and (5) as far as the order of ϕ and ψ is concerned. Beginners often encounter considerable difficulty in appreciating the distinction between *if* and *only if*.

Likewise, the use of the word *necessary* in (7) often causes confusion. Notice that to say that ψ is a necessary condition for ϕ does not mean that ψ on its own is enough to guarantee ϕ. Rather what it says is that ψ will have to hold *before there can even be any question of ϕ holding*.

The following diagram might help you remember the distinction between *necessary* and *sufficient*:

ϕ	IMPLIES	ψ
\uparrow		\uparrow
sufficient		necessary

(Think of the word "sun". This will remind you of the order.)

Because equivalence reduces to implication both ways, it follows from the above discussion that the following are also equivalent:

(1) ϕ is equivalent to ψ

(2) ϕ is necessary and sufficient for ψ

(3) ϕ if and only if ψ

A common abbreviation for the phrase *if and only if* is *iff* (or occasionally *iffi*). Thus we often write

$$\phi \text{ iff } \psi$$

Note that if we were to be strict about the matter, the above discussion of equivalent terminologies refers to implication and equivalence, not their formal counterparts the conditional and the biconditional. However, mathematicians frequently use the symbol \Rightarrow as an abbreviation for *implies* and \Leftrightarrow as an abbreviation for *is equivalent to*, so the different terminologies often are used together with these formally defined symbols.

Although this is invariably confusing to beginners, it's simply the way mathematical practice has evolved, so there is no getting around it. You would be entirely

justified in throwing your hands up at what seems on the face of it to be sloppy practice. After all, if there are genuine problems with the meanings of certain words that necessitate a lengthy discussion and the formulation of formal definitions of concepts that are not identical to their everyday counterparts (such as the difference between the conditional and implication), why do mathematicians then promptly revert to the everyday notions that they began by observing to be problematic?

Here is why the professionals do this. The conditional and biconditional only differ from implication and equivalence in situations *that do not arise in the course of normal mathematical practice*. In any real mathematical context, the conditional "is" implication and the biconditional "is" equivalence. Thus, having made note of where the formal notions differ from the everyday ones, mathematicians simply move on and turn their attention to other things.

Truth Tables

Since we have defined the logical connectives $\wedge, \vee, \neg, \Rightarrow, \Leftrightarrow$ in terms of truth and falsity alone, and not to meaning, it is possible to represent (or illustrate) them by means of a table: a *truth table*. We introduce two symbols: T to denote "true" and F to denote "false". The behavior/definition of $\phi \wedge \psi$ can then be illustrated by the table:

ϕ	ψ	$\phi \wedge \psi$
T	T	T
T	F	F
F	T	F
F	F	F

In the first two columns appear all the possible combinations of values of T and F that the two statements ϕ and ψ can have. In the third column we give the value $\phi \wedge \psi$ achieves according to each assignment of T or F to ϕ and ψ. Thus, we see that $\phi \wedge \psi$ is T only when both ϕ and ψ are T.

For $\phi \vee \psi$ we have the table

ϕ	ψ	$\phi \vee \psi$
T	T	T
T	F	T
F	T	T
F	F	F

Again, the definition of $\neg \phi$ can be represented thus:

ϕ	$\neg \phi$
T	F
F	T

For $\phi \Rightarrow \psi$ we have:

ϕ	ψ	$\phi \Rightarrow \psi$
T	T	T
T	F	F
F	T	T
F	F	T

One can go on to construct truth tables for more complicated expressions. Consider, for example, the expression $(\phi \wedge \psi) \vee (\neg\phi)$. We can build its table column by column as follows:

ϕ	ψ	$\phi \wedge \psi$	$\neg\phi$	$(\phi \wedge \psi) \vee (\neg\phi)$
T	T	T	F	T
T	F	F	F	F
F	T	F	T	T
F	F	F	T	T

We can also draw up tables for expressions such as $(\phi \wedge \psi) \vee \theta$, but if there are n constituent statements involved there will be 2^n rows in the table, so already $(\phi \wedge \psi) \vee \theta$ needs 8 rows!

Truth tables can be useful in checking that two rather complex statements are equivalent. For, by our definition of equivalence, two statements will be equivalent if they have the same truth table. For example, we can demonstrate the equivalence of

$$\neg(\phi \wedge \psi) \quad \text{and} \quad (\neg\phi) \vee (\neg\psi)$$

as follows:

ϕ	ψ	$\phi \wedge \psi$	$\neg(\phi \wedge \psi)$ *	$\neg\phi$	$\neg\psi$	$(\neg\phi) \vee (\neg\psi)$ *
T	T	T	F	F	F	F
T	F	F	T	F	T	T
F	T	F	T	T	F	T
F	F	F	T	T	T	T

Since the two columns marked * are identical, we know that the two expressions are equivalent.

Contrapositives

The *contrapositive* of a conditional $\phi \Rightarrow \psi$ is the conditional

$$\neg\psi \Rightarrow \neg\phi$$

(Note that the introduction of the negation sign is accompanied by a change in the direction of the arrow.)

For example, for the implication

If $2^n - 1$ is prime, then n is prime

the contrapositive implication is

If n is composite (i.e., not prime), then $2^n - 1$ is composite

The following result is the logical basis for the mathematical concept of proof by contrapositive, where an implication is proved by establishing its contrapositive.

THEOREM 2.2.1
Any conditional is equivalent to its contrapositive.

PROOF This is a straightforward exercise in constructing truth tables. ☐

By virtue of this theorem, in order to prove that $2^n - 1$ can be prime only if n is prime, a legitimate approach is to prove that if n is composite then $2^n - 1$ is composite. (Exercise: Give such a proof.)

The contrapositive of a conditional should not be confused with the *converse*. The *converse* of $\phi \Rightarrow \psi$ is the conditional $\psi \Rightarrow \phi$. There is in general no connection between the truth value of a conditional and its converse.

Exercises 2.2

(1) By a *denial* of a statement ϕ we mean any statement equivalent to $\neg\phi$. Give a useful denial of each of the following statements.

 (a) 34,159 is a prime number.

 (b) Roses are red and violets are blue.

 (c) If there are no hamburgers, I'll have a hot dog.

 (d) Fred will go but he will not play.

 (e) The number x is either negative or greater than 10.

 (f) We will win the first game or the second.

(2) Which of the following conditions is *necessary* for the natural number n to be divisible by 6?

 (a) n is divisible by 3.

 (b) n is divisible by 9.

 (c) n is divisible by 12.

 (d) $n = 24$.

 (e) n^2 is divisible by 3.

 (f) n is even and divisible by 3.

(3) In (2), which conditions are *sufficient* for n to be divisible by 6?

(4) In (2), which conditions are *necessary and sufficient* for n to be divisible by 6?

(5) Let m, n denote any two natural numbers. Prove that mn is odd iff m and n are odd.

(6) With reference to (5), is it true that mn is even iff m and n are even?

(7) Show that $\phi \Leftrightarrow \psi$ is equivalent to $(\neg\phi) \Leftrightarrow (\neg\psi)$. How does this relate to your answers to questions (5) and (6) above?

(8) Construct truth tables to illustrate the following:

 (a) $\phi \Leftrightarrow \psi$

 (b) $\phi \Rightarrow (\psi \vee \theta)$

(9) Use truth tables to prove that the following are equivalent:

 (a) $\neg(\phi \Rightarrow \psi)$ and $\phi \wedge (\neg\psi)$

 (b) $\phi \Rightarrow (\psi \wedge \theta)$ and $(\phi \Rightarrow \psi) \wedge (\phi \Rightarrow \theta)$

 (c) $(\phi \vee \psi) \Rightarrow \theta$ and $(\phi \Rightarrow \theta) \wedge (\psi \Rightarrow \theta)$

(10) Verify the equivalences in (b) and (c) in the previous question by means of a logical argument. (So, in the case of (b), for example, you must show that assuming ϕ and deducing $\psi \wedge \theta$ is the same as both deducing ψ from ϕ and θ from ϕ.)

(11) Use truth tables to prove the equivalence of $\phi \Rightarrow \psi$ and $(\neg\psi) \Rightarrow (\neg\phi)$

(12) Use truth tables to show that $\phi \Rightarrow \psi$ and $\psi \Rightarrow \phi$ are not equivalent. Give an example of statements ϕ and ψ such that $\phi \Rightarrow \psi$ is true but $\psi \Rightarrow \phi$ is false.

(13) Express the connective

$$\phi \text{ unless } \psi$$

in terms of the standard propositional connectives.

(14) Identify the antecedent and the consequent in each of the following conditionals:

(a) If apples are red then oranges are green.

(b) The differentiability of a function f is sufficient for f to be continuous.

(c) A function f is bounded if f is integrable.

(d) A sequence s is bounded whenever s is convergent.

(e) It is necessary that n is prime in order for $2^n - 1$ to be prime.

(f) The team wins only when Karl is playing.

(g) When Karl plays the team wins.

(h) The team wins when Karl plays.

(15) Write the converse and contrapositive of each conditional in the previous question.

(16) Let $\dot{\vee}$ denote the "exclusive or" that corresponds to the English expression "either one or the other but not both". Construct a truth table for this connective.

(17) Which of the following pairs of propositionals are equivalent?

(a) $\neg(P \vee Q), \neg P \wedge \neg Q$

(b) $\neg P \vee \neg Q, \neg(P \vee \neg Q)$

(c) $\neg(P \wedge Q), \neg P \vee \neg Q$

(d) $\neg(P \Rightarrow (Q \wedge R)), \neg(P \Rightarrow Q) \vee \neg(P \Rightarrow R)$

(e) $P \Rightarrow (Q \Rightarrow R), (P \wedge Q) \Rightarrow R$

(18) Prove Theorem 2.2.1.

(19) Give, if possible, an example of a true conditional sentence for which

(a) the converse is true.

(b) the converse is false.

(c) the contrapositive is true.

(d) the contrapositive is false.

(20) Imagine that a psychologist lays four cards on a table in front of you. He tells you that each card has a number on one side and a letter on the other. On the uppermost faces of the cards you see the four symbols

$$\text{E} \quad \text{K} \quad 4 \quad 7$$

The psychologist then tells you that the cards are printed according to the rule: If a card has a vowel on one side, it has an even number on the other side. He then asks you which cards do you *have* to turn over to be sure that all four cards satisfy this rule. How do you answer? [This problem was first posed by the British psychologist Peter Wason in the early 1970s. Most people get it wrong.]

(21) You are in charge of a party where there are young people. Some are drinking alcohol, others soft drinks. Some are old enough to drink alcohol legally, others are under age. You are responsible for ensuring that the drinking laws are not broken, so you have asked each person to put his or her photo ID on the table. At one table are four young people. One person has a beer, another has a Coke, but their IDs happen to be face down so you can't see their ages. You can, however, see the IDs of the other two people. One is under the drinking age, the other is above it. Unfortunately, you are not sure if they are drinking 7-Up or vodka and tonic. Which IDs and/or drinks do you need to check to make sure that no one is breaking the law?

(22) Compare the logical structures of the two previous questions and comment on your answers to those two questions. In particular, identify any logical rules you used in solving each problem, say which one was easier, and why you felt it was easier.

2.3 The Language of Mathematics: Part 2

And so to the final part of our investigation of language. We introduce the two *quantifiers*:

there exists, for all

In mathematics we frequently make *existence* assertions. For example

the equation $x^2 + 2x + 1 = 0$ has a real root

expresses an existence claim. The existential nature of this assertion can be made more explicit by rewriting it in the form:

there exists a real number x such that $x^2 + 2x + 1 = 0$

Or again

$\sqrt{2}$ is rational

also expresses an existence claim, though it certainly does not look much like an existence statement until we write it in the form

there exist natural numbers p and q such that $\sqrt{2} = p/q$

The first example here is a true statement (take $a = -1$); the second is false (as we prove later). Incidentally, the second example is not unique in having its existential nature hidden: there are many mathematical statements that are really existence assertions, but that do not appear so on the surface.

We use the symbol

$$\exists x$$

to mean

there exists an x such that ...

Thus the above two examples may be written symbolically as:

- $\exists x[x^2 + 2x + 1 = 0]$
- $\exists p \exists q[\sqrt{2} = p/q]$

But there is a potential problem with this notation. In the original English-language sentences we specified the kinds of objects involved in these quantifiers, real numbers in the first case and natural numbers in the second. But this is lost in the symbolic versions, where we rely upon the context or the reader's experience (or both) to determine that the x is a real number variable and the p, q are variables for natural numbers.

Now, in point of fact, we often can (and do) simply rely on context, aided perhaps by the fairly loose conventions that x and y usually denote real numbers, p, q, m, n denote whole numbers, and z, w denote complex numbers. (But do note that these conventions are *very* loose, and should not be relied upon without some additional indications of the usage.)

Another solution is to modify the quantifier notation so as to specify the kind of objects being quantified. For instance, we could make use of the symbol \mathcal{R} for the set of real numbers (see Chapter 3) and write

$$(\exists x \in \mathcal{R})$$

to mean

there exists a real number such that ...

("$x \in \mathcal{R}$" means "x is a member of \mathcal{R}". See Chapter 3 for a discussion of basic set theory.)

Thus, to say that the equation $x^2 + 2x + 1 = 0$ has a real root we could write

$$(\exists x \in \mathcal{R})(x^2 + 2x + 1 = 0)$$

Again, using the standard notation \mathcal{N} for the set of natural numbers, we could write

$$(\exists p \in \mathcal{N})(\exists q \in \mathcal{N})[\sqrt{2} = p/q]$$

to say that $\sqrt{2}$ is rational.

To avoid proliferation of notation, where there is a repetition of a quantifier as here, we could (and often do) abbreviate this to

$$(\exists p, q \in \mathcal{N})(\sqrt{2} = p/q)$$

The back-to-front E that we use for the "exists" symbol comes from the word "exists".

As well as saying that certain objects exist, we also have frequent need to say that something holds *for all x*. We use the symbol

$$\forall x$$

to mean

for all x it is the case that ...

Again, if we wish to specify what sort of x we are considering, we can modify the notation, writing, for instance,

$$(\forall x \in \mathcal{R})$$

to mean

for all real numbers x it is the case that ...

and

$$(\forall x \in \mathcal{N})$$

to mean

for all natural numbers x it is the case that ...

For example, to say that $\sqrt{2}$ is irrational we can write

$$(\forall p \in \mathcal{N})(\forall q \in \mathcal{N})(\sqrt{2} \neq p/q)$$

Notice that the symbol \forall is just an upside-down A, coming from the word "all".

The symbol \exists is called the *existential quantifier*; \forall is the *universal quantifier*.

Most statements in mathematics involve combinations of both kinds of quantifier. For instance, to make the assertion that there is no largest natural number needs two quantifiers, thus:

$$(\forall m \in \mathcal{N})(\exists n \in \mathcal{N})(n > m)$$

This reads: for all natural numbers m it is the case that there exists a natural number n such that n is greater than m.

Notice that the order in which quantifiers appear can be of paramount importance. For example, if we switch the order in the above we get

$$(\exists n \in \mathcal{N})(\forall m \in \mathcal{N})(n > m)$$

This asserts that there is a natural number which exceeds all natural numbers — an assertion that is clearly false!

Exercises 2.3

(1) Express the following as existence assertions:

 (a) The equation $x^3 = 27$ has a natural number solution.

 (b) 1,000,000 is not the largest natural number.

 (c) The natural number n is not a prime.

(2) Express the following as "for all" assertions:

 (a) The equation $x^3 = 28$ does not have a natural number solution.

 (b) 0 is less than every natural number.

 (c) The natural number n is a prime.

(3) Express the following in symbolic form using quantifiers for people:

 (a) Everybody loves somebody.

 (b) Everyone is tall or everyone is short.

 (c) Everone is tall or short.

 (d) Nobody is at home.

 (e) If a man comes, all the women will leave.

(4) Express the following using quantifiers. (In each case your quantifiers may refer only to the sets \mathcal{R} and \mathcal{N}.)

 (a) The equation $x^2 + a = 0$ has a real root for any real number a.

 (b) The equation $x^2 + a = 0$ has a real root for any negative real number a.

 (c) Every real number is rational.

 (d) There is an irrational number.

 (e) There is no largest irrational number. (This one looks quite complex.)

(5) Let C be the set of all cars, let $D(x)$ mean that x is domestic, and let $M(x)$ mean that x is badly made. Express the following in symbolic form using these symbols:

 (a) All domestic cars are badly made.

 (b) All foreign cars are badly made.

 (c) All badly made cars are domestic.

 (d) There is a domestic car that is not badly made.

 (e) There is a foreign car that is badly made.

(6) Express the following sentence using quantifiers for real numbers:

 There is a rational number between any two unequal real numbers.

(Use only logical connectives and quantifiers, the order relation <, and the symbol $Q(x)$ having the meaning "x is rational".)

(7) Express the following famous statement (by Abraham Lincoln) using quantifiers for people and times: "You can fool some of the people all of the time and all of the people some of the time, but you cannot fool all of the people all of the time."

2.4 Properties of Quantification

We examine the way quantifiers behave and how they interact with the logical connectives we introduced earlier in the chapter.

We start by taking up the issue raised already, as to what kind of objects are denoted by the variable in a quantifier.

Domain of Quantification

Associated with any use of quantifiers there is what is known as a *domain of quantification*: the collection of all objects that the quantifiers refer to. This may be the collection of all real numbers, the collection of all natural numbers, the collection of all complex numbers, or some other collection. In many cases, the overall context determines the domain. For instance, if we are studying real analysis, then unless otherwise mentioned it may safely be assumed that any quantifier refers to the real numbers. But on occasions there is no alternative but to be quite explicit as to what is the domain under discussion.

To illustrate how it can sometimes be important to specify the domain, consider the two mathematical statements

$$\forall x (x > 2) \quad \text{and} \quad \forall x (x \geq 3)$$

These are equivalent with respect to the domain \mathcal{Z} of all integers but not equivalent when the domain is \mathcal{R}.

At the risk of confusing you, I should mention that, in practice, mathematicians often omit mention of universal quantification, writing expressions such as

$$x \geq 0 \Rightarrow \sqrt{x} \geq 0$$

when what is meant is

$$(\forall x \in \mathcal{R})[x \geq 0 \Rightarrow \sqrt{x} \geq 0]$$

This is known as *implicit quantification*. Although I do not use this convention in this book, implicit quantification is fairly common, so you should be aware of it.

Care has to be exercised when quantifiers are combined with the logical connectives ∧, ∨, etc.

As an illustration of the various pitfalls that can arise, suppose the domain under discussion is the set of natural numbers. Let $E(x)$ be the statement "x is even" and let $O(x)$ be the statement "x is odd".

The statement

$$\forall x[E(x) \vee O(x)]$$

says that for every natural number x, x is either even or odd (or both). This is clearly true.

On the other hand, the statement

$$\forall x\, E(x) \vee \forall x\, O(x)$$

is false, since it asserts that either all natural numbers are even or else all natural numbers are odd (or both), whereas in fact neither of these alternatives is the case.

Thus, in general you cannot "move a $\forall x$ inside brackets." More precisely, if you do, you can end up with a very different statement, not equivalent to the original one.

Again, the statement

$$\exists x[E(x) \wedge O(x)]$$

is false, since it claims that there is a natural number that is both even and odd, whereas the statement

$$\exists x\, E(x) \wedge \exists x\, O(x)$$

claims that there is a natural number that is even and there is a natural number that is odd, which is true.

Thus "moving a $\exists x$ inside brackets" can also lead to a statement that is not equivalent to the original one.

Notice that although the last statement above uses the same variable x in both parts of the conjunction, the two conjuncts operate separately.

You should make sure you fully appreciate the distinction between *all* the above example statements involving quantifiers.

Subdomains

Very often, in the course of an argument, we use quantifers that are restricted to a smaller collection than the original domain. For example, in real analysis (where the unspecified domain is usually the set \mathcal{R} of all real numbers) we often need to talk about "all positive numbers" or "all negative numbers", and in number theory (where the unspecified domain is the set \mathcal{N} of all natural numbers) we have quantifiers such as "for all prime numbers", etc.

One way to handle this has been done already. We can modify the quantifier notation, allowing quantifiers of the form

$$(\forall x \in A), \quad (\exists x \in A)$$

where A is some subcollection of the domain.

Another way is to specify the objects being quantified within the non-quantifier part of the formula. For example, suppose the domain under discussion is the set of all animals. Thus, any variable x is assumed to denote an animal. Let $L(x)$ mean that "x is a leopard" and let $S(x)$ mean that "x has spots". Then the sentence "all leopards have spots" can be written like this:

$$\forall x[L(x) \Rightarrow S(x)]$$

In English, this reads literally as: "For all animals x, if x is a leopard *then* x has spots." This is rather cumbersome English, but the mathematical version turns out to be far more convenient than the alternative of using a modified quantifier of the form $(\forall x \in \mathcal{L})$ where \mathcal{L} denotes the set of all leopards.

Beginners often make the mistake of rendering the original sentence "all leopards have spots" as

$$\forall x[L(x) \wedge S(x)]$$

In English, what this says is: "For all animals x, x is both a leopard and has spots." Or, smoothing out the English a bit, "All animals are leopards and have spots." This is obviously false; not all animals are leopards, for one thing.

Part of the reason for the confusion is probably the fact that the mathematics goes differently in the case of existential sentences. For example, consider the sentence "There is a horse that has spots." If we let $H(x)$ mean that "x is a horse", then this sentence translates into the mathematical sentence

$$\exists x[H(x) \wedge S(x)]$$

Literally: "There is an animal that is both a horse *and* has spots."

Contrast this with the sentence

$$\exists x[H(x) \Rightarrow S(x)]$$

This says that "There is an animal such that *if* it is a horse, *then* it has spots." This does not seem to say anything much, and is certainly not at all the same as saying that there is a spotted horse.

In symbolic terms, the modified quantifer notation

$$(\forall x \in \mathcal{A})\phi(x)$$

may be regarded as an abbreviation for the expression

$$\forall x[A(x) \Rightarrow \phi(x)]$$

where $A(x)$ is the property of x being in the collection \mathcal{A}.

Likewise, the notation

$$(\exists x \in \mathcal{A})\phi(x)$$

may be regarded as an abbreviation for

$$\exists x[A(x) \wedge \phi(x)]$$

Negation and Quantification

In real analysis and in many other parts of mathematics, it is important to be able to negate statements involving quantifiers (and end up with the correct answer). Of course, one can negate any statement by simply putting a negation symbol in front. But often what we want is a *positive* assertion, not a *negative* one. What is meant by "positive" here should become clear from our examples. Roughly speaking, a positive statement is one containing no negation symbol, or else one in which the negation symbols are as far inside the statement as is possible without the resulting expression being unduly cumbersome.

Let $A(x)$ denote some property of x. (For example, $A(x)$ could say that x is a real root of the equation $x^2 + 2x + 1 = 0$.)

Suppose it is not the case that $\forall x\, A(x)$ is true. That is, assume

$$\neg[\forall x\, A(x)]$$

Then, if it is not the case that all x satisfy $A(x)$, what must happen is that at least one of the x must fail to satisfy $A(x)$.

In other words, for at least one x, $\neg A(x)$ must be true. In symbols, this can be written as

$$\exists x[\neg A(x)]$$

Hence $\neg[\forall x\, A(x)]$ implies $\exists x[\neg A(x)]$.

Now suppose

$$\exists x[\neg A(x)]$$

Thus there will be an x for which $A(x)$ fails. Hence $A(x)$ does not hold for every x (it fails for the x where it fails!).

In other words, it is false that $A(x)$ holds for all x. In symbols,

$$\neg[\forall x\, A(x)]$$

Thus $\exists x[\neg A(x)]$ implies $\neg[\forall x\, A(x)]$.

Taken together, the two implications just established produce the equivalence

$$\neg[\forall x\, A(x)] \text{ if and only if } \exists x[\neg A(x)]$$

A similar argument shows that

$$\neg[\exists x\, A(x)] \text{ if and only if } \forall x[\neg A(x)]$$

(Exercise: Provide this argument.)

Now let us return to that earlier problem about domestic cars, where we want to negate the statement

All domestic cars are badly made

Let us formulate this symbolically using the notation of Exercise 2.3(5). If you got part (a) of this question correct, you should have the formulation

$$(\forall x \in C)[D(x) \Rightarrow M(x)]$$

Negating this gives

$$(\exists x \in C)\neg[D(x) \Rightarrow M(x)]$$

(One common cause of confusion. Why do we not say $(\exists x \notin C)$? The answer is that the '$\in C$' part simply tells us which kind of x we are to consider. Since our original statement concerns domestic cars, so will its negation.)

Consider now the part

$$\neg[D(x) \Rightarrow M(x)]$$

We have seen already that this is equivalent to

$$D(x) \wedge (\neg M(x))$$

Hence for our negated statement (in positive form now) we get

$$(\exists x \in C)[D(x) \wedge (\neg M(x))]$$

In words, there is a car that is domestic and is not badly made; i.e., there is a domestic car that is not badly made.

We can also obtain this result directly as follows, without going through the above symbolic manipulations.

If it is not the case that all domestic cars are badly made, then it must be the case that at least one of them fails to be badly made. Hence, as this argument reverses, the required negation is that at least one domestic car is not badly made.

The issue discussed above causes problems for enough beginners to warrant some further examples.

Suppose the universe is \mathcal{N} and that $P(x)$ denotes the property "x is a prime" and $O(x)$ the property "x is odd". Consider the sentence

$$\forall x[P(x) \Rightarrow O(x)]$$

This says that all primes are odd, which is false. Then the negation of this sentence will have the (positive) form

$$\exists x[P(x) \wedge \neg O(x)]$$

Thus the \forall becomes a \exists and the \Rightarrow becomes a \wedge. In words, the negation reads "There is a prime that is not odd", or more colloquially, "there is an even prime". This is, of course, true.

If we modify the original sentence to read

$$(\forall x > 2)[P(x) \Rightarrow O(x)]$$

(i.e., all primes greater than 2 are odd, which is true) then the negation of this sentence can be written as

$$(\exists x > 2)[P(x) \wedge \neg O(x)]$$

(i.e., there is an even prime bigger than 2) which is false.

One thing to notice about this example is that the quantifier $(\forall x > 2)$ changes to $(\exists x > 2)$, *not* to $(\exists x < 2)$. Likewise, negation of the quantifier $(\exists x > 2)$ leads to $(\forall x > 2)$, *not* to $(\forall x < 2)$. You should make sure you understand the reason for this behavior.

Again, suppose that the universe were the set of all people, and let $P(x)$ be the property of being a player for a certain sports team and $H(x)$ the property of being healthy. Then the sentence

$$\exists x[P(x) \wedge \neg H(x)]$$

expresses the claim that there is an unhealthy player. Negating this gives

$$\forall x[\neg P(x) \vee H(x)]$$

This is a bit unnatural to read in English, but by virtue of the way we defined \Rightarrow, it can be rewritten as

$$\forall x[P(x) \Rightarrow H(x)]$$

and this has the natural reading that "all players are healthy".

Finally, let the universe be the set, \mathcal{Q}, of all rationals, and consider the sentence

$$\forall x[x > 0 \Rightarrow \exists y(xy = 1)]$$

This says that every positive rational has a multiplicative inverse (which is true). The negation of this sentence works out as follows.

$$\neg \forall x[x > 0 \Rightarrow \exists y(xy = 1)] \quad \text{iff } \exists x[x > 0 \wedge \neg \exists y(xy = 1)]$$

$$\text{iff } \exists x[x > 0 \wedge \forall y(xy \neq 1)]$$

In words, there is a positive rational with the property that no y exists such that $xy = 1$, i.e., there is a positive rational with no multiplicative inverse.

In order to negate statements with more than one quantifier, the idea is to start at the outside and work inwards, handling each quantifier in turn. The overall effect being that the negation symbol moves inwards, changing each \forall to an \exists and each \exists to a \forall as it passes. Thus, for example

$$\neg[\forall x \exists y \forall z A(x, y, z)] \quad \text{iff } \exists x \neg[\exists y \forall z A(x, y, z)]$$

$$\text{iff } \exists x \forall y \neg[\forall z A(x, y, z)]$$

$$\text{iff } \exists x \forall y \exists z \neg[A(x, y, z)]$$

Unique Existence

One further quantifier that is often useful is

"there exists a unique x such that ..."

The usual notation for this quantifier is

$$\exists!$$

This quantifier can be defined in terms of the other quantifiers, by taking

$$\exists!x\phi(x)$$

to be an abbreviation for

$$\exists x\phi(x) \wedge \forall y[\phi(y) \Rightarrow x = y]$$

(Make sure you understand why this last formula expresses unique existence.)

Exercises 2.4

(1) Translate the following sentences into symbolic form using quantifiers. In each case the assumed domain is given in parentheses.

 (a) All students like pizza. (All people)

 (b) One of my friends does not have a car. (All people)

 (c) Some elephants do not like muffins. (All animals)

 (d) Every triangle is isosceles. (All geometric figures)

 (e) Some of the students in the class are not here today. (All people)

 (f) Everyone loves somebody. (All people)

 (g) If a man comes, all the women will leave. (All people)

 (h) All people are tall or short. (All people)

 (i) All people are tall or all people are short. (All people)

 (j) Not all precious stones are beautiful. (All stones)

 (k) Nobody loves me. (All people)

 (l) At least one British snake is poisonous. (All snakes)

 (m) At least one British snake is poisonous. (All animals)

(2) Which of the following are true? The domain for each is given in parentheses.

 (a) $\forall x(x + 1 \geq x)$ (Real numbers)

 (b) $\exists x(2x + 3 = 5x + 1)$ (Natural numbers)

 (c) $\exists x(x^2 + 1 = 2^x)$ (Real numbers)

 (d) $\exists x(x^2 = 2)$ (Rational numbers)

 (e) $\exists x(x^2 = 2)$ (Real numbers)

 (f) $\forall x(x^3 + 17x^2 + 6x + 100 \geq 0)$ (Real numbers)

 (g) $\exists x(x^3 + x^2 + x + 1 \geq 0)$ (Real numbers)

 (h) $\forall x \exists y(x + y = 0)$ (Real numbers)

 (i) $\exists x \forall y(x + y = 0)$ (Real numbers)

 (j) $\forall x \exists! y(y = x^2)$ (Real numbers)

 (k) $\forall x \exists! y(y = x^2)$ (Natural numbers)

 (l) $\forall x \exists y \forall z(xy = xz)$ (Real numbers)

 (m) $\forall x \exists y \forall z(xy = xz)$ (Prime numbers)

 (n) $\forall x \exists y(x \geq 0 \Rightarrow y^2 = x)$ (Real numbers)

 (o) $\forall x[x < 0 \Rightarrow \exists y(y^2 = x)]$ (Real numbers)

 (p) $\forall x[x < 0 \Rightarrow \exists y(y^2 = x)]$ (Positive real numbers)

(3) Negate each of the mathematical statements you wrote in question (1), putting your answers in positive form. Express each negation in natural, idiomatic English.

(4) Negate each of the statements in question (2), putting your answers in positive form.

(5) Negate the following statements and put each answer into positive form:

 (a) $(\forall x \in \mathcal{N})(\exists y \in \mathcal{N})(x + y = 1)$

 (b) $(\forall x > 0)(\exists y < 0)(x + y = 0)$ (where x, y are real number variables)

 (c) $\exists x(\forall \epsilon > 0)(-\epsilon < x < \epsilon)$ (where x, y are real number variables)

 (d) $(\forall x \in \mathcal{N})(\forall y \in \mathcal{N})(\exists z \in \mathcal{N})(x + y = z^2)$

(6) The standard definition of a real function f being *continuous at a point* $x = a$ is

$$(\forall \epsilon > 0)(\exists \delta > 0)(\forall x)[|x - a| < \delta \Rightarrow |f(x) - f(a)| < \epsilon]$$

Write down the formal definition for f being *discontinuous at a*.

(7) Give a negation (in positive form) of the sentence which you met in Exercises 2.3(7): "You can fool some of the people all of the time and all of the people some of the time, but you cannot fool all of the people all of the time."

2.5 Proofs in Mathematics

A *proof* of a statement in mathematics is a logically sound argument that establishes the truth of the statement. There are many types of proof. Our purpose here is just to mention a few of these and see how they relate to our discussion of language.

Proving a Conditional

Suppose first we wish to establish the truth of a conditional

$$\phi \Rightarrow \psi$$

Well, since (by our definition of the conditional) this will certainly be true whenever ϕ is false, we need only consider the case when ϕ is true. That is, we can *assume* ϕ. For the conditional to be valid now, ψ must also be true. Thus, using our assumption that ϕ is true we must now present an argument that demonstrates the truth of ψ. This, of course, accords with our everyday understanding of implication. Thus, when it comes to *proving* conditionals, the problems concerning the distinction between the conditional and real implication that we encountered in Section 2.1 do not arise.

For example, suppose we are given a pair x, y of real numbers. We do not know just which numbers they are, but nevertheless we want to prove the statement

$$(x \text{ and } y \text{ are rational numbers}) \Rightarrow (x + y \text{ is a rational number})$$

We start by *assuming* x and y are rational numbers. Then we can find integers p, q, m, n such that $x = p/m$, $y = q/n$. Then

$$x + y = \frac{p}{m} + \frac{q}{n} = \frac{pn + qm}{mn}$$

Hence, as $pn + qm$ and mn are integers, we conclude that $x + y$ is rational. The statement is proved.

Conditionals involving quantifiers are often best handled by using the equivalence of $\phi \Rightarrow \psi$ with the contrapositive $(\neg\psi) \Rightarrow (\neg\phi)$.

Suppose, for instance, that for some given, unknown angle θ we wish to prove the conditional

$$(\sin\theta \neq 0) \Rightarrow (\forall n \in \mathcal{N})(\theta \neq n\pi)$$

This statement is equivalent to

$$\neg(\forall n \in \mathcal{N})(\theta \neq n\pi) \Rightarrow \neg(\sin\theta \neq 0)$$

which reduces to the positive form

$$(\exists n \in \mathcal{N})(\theta = n\pi) \Rightarrow (\sin\theta = 0)$$

This is an implication we know to be correct. This proves the original implication by virtue of what equivalence means. For in order to prove a statement it is enough to prove any equivalent statement. In the above case the proof was an example of what is sometimes referred to as *proof by contrapositive*.

Proof by Contradiction

Another common method of proof is the so-called method of *proof by contradiction (reductio ad absurdum)*. This is related to the method of proof by contrapositive mentioned above. We wish to prove some statement ϕ but cannot see how

to begin. So we *assume* $\neg\phi$ and proceed to deduce some obviously false statement (e.g., $0 = 1$). Since our conclusion is false it follows (assuming our reasoning is sound) that our initial assumption of $\neg\phi$ has to be false. Hence ϕ is shown to be true.

This relates to the method of proof by contrapositive as follows. Let F denote any false statement, T any true statement. We *assume* $\neg\phi$ and *deduce* F. So we *prove* the conditional

$$(\neg\phi) \Rightarrow F$$

But this is logically equivalent to the contrapositive

$$(\neg F) \Rightarrow (\neg(\neg\phi))$$

i.e., to

$$T \Rightarrow \phi$$

Hence our proof also establishes the conditional $T \Rightarrow \phi$. But T is true, so the truth of the conditional means that ϕ is true, as required.

An excellent illustration of the method of proof by contradiction is furnished by the following result, promised earlier.

THEOREM 2.5.1
$\sqrt{2}$ *is irrational.*

PROOF Assume, on the contrary, that $\sqrt{2}$ were rational. Then we could find natural numbers p and q such that

$$\sqrt{2} = p/q$$

where p and q have no common factors. Squaring gives

$$2 = p^2/q^2$$

which rearranges to give

$$p^2 = 2q^2$$

Thus p^2 is even. Hence p must be even, since $\text{odd}^2 = \text{odd}$. Thus there is a natural number r such that $p = 2r$. Substituting for p in the last equation gives

$$4r^2 = 2q^2$$

and canceling 2 gives

$$2r^2 = q^2$$

Thus q^2 is even. Hence q must be even. But p is even and p and q have no common factors, so we have a contradiction. Hence our original assumption that $\sqrt{2}$ was rational must be false. In other words, $\sqrt{2}$ must be irrational, which is what we set out to prove. ▯

Proving Existence Statements

Proving existence assertions is often straightforward. In order to prove the statement

$$\exists x \, A(x)$$

it is enough to find one particular x that satisfies $A(x)$. For instance, if we wish to prove that

there exists an irrational number

it suffices to quote the above theorem. Not only does this tell us that an irrational number exists, it provides us with a specific example.

However, it is not always the case that an existence proof identifies a particular object that makes the existence assertion true. There are many instances in mathematics where one knows that, say, a real number exists that satisfies a certain property, but one has no idea what such an x looks like, or even whether it is positive or negative. Indeed, advanced mathematics abounds with examples. We content ourselves here with one fairly simple example.

THEOREM 2.5.2
There are irrational numbers a and b such that a^b is rational.

PROOF By Theorem 2.5.1 we know that $\sqrt{2}$ is irrational. Consider the number $r = \sqrt{2}^{\sqrt{2}}$. If r is rational, then we can take $a = b = \sqrt{2}$ to establish the theorem. If, on the other hand, r is irrational, then $a = r$ and $b = \sqrt{2}$ satisfy the theorem, since

$$a^b = (\sqrt{2}^{\sqrt{2}})^{\sqrt{2}} = \sqrt{2}^2 = 2 \qquad \qquad \Box$$

The thing to notice about the above proof is that it does not tell us which particular pair a, b of numbers satisfies the requirements of the theorem.[3]

Of course, in the above proof we do know that the actual pair is one of two possible pairs. Many examples that occur naturally in more advanced parts of mathematics do not even provide a finite range of possibilities for a solution: they simply yield the conclusion that there is such-and-such an object. Such proofs are often referred to as *non-constructive proofs*. Quite often they involve proof by contradiction.

Proving "For All" Statements

Finally, we look at how to prove a statement of the form

$$\forall x \, A(x)$$

[3]In fact a considerably more sophisticated proof shows that the number r is indeed irrational, so the second of the two pairs considered is the one that works. But the proof given here does not establish this fact.

One possibility is to take an arbitrary x and show that it must satisfy $A(x)$. For instance, suppose we wish to prove the assertion

$$(\forall n \in \mathcal{N})(\exists m \in \mathcal{N})(m > n^2)$$

We can do this as follows.

Let n be an arbitrary natural number. Then n^2 is a natural number. Hence $m = n^2 + 1$ is a natural number. Since $m > n^2$, this shows that

$$(\exists m \in \mathcal{N})(m > n^2)$$

This is a proof because our original n was quite *arbitrary*. We said nothing at all about n: it could be any natural number. Hence the argument is valid for *all* n in \mathcal{N}. This is not the same as picking a *particular* n. If we had randomly chosen, say, $n = 37$, the proof would not have been valid — even though we had chosen this n quite at random. For instance, suppose we wanted to prove

$$(\forall n \in \mathcal{N})(n^2 = 81)$$

By picking at random a particular n we might happen to pick $n = 9$. But this does not prove the statement of course, because our choice was an arbitrary choice (albeit an unlucky one as far as our goal is concerned) of a *particular* n, and not a choice of an *arbitrary* n.

In practice what this amounts to is that whenever we start a proof by saying "let n be arbitrary," we use the symbol n throughout the proof, and assume also that the value of n remains constant throughout, but we make absolutely no restriction on what the value of n is.

There are other possibilities for proving universally quantified statements. In particular, statements of the form

$$(\forall n \in \mathcal{N})A(n)$$

where the quantification is over all natural numbers, are often proved by a method known as *induction*.

Proof by Induction

To prove a statement of the form

$$(\forall n \in \mathcal{N})A(n)$$

by induction, you verify the following two statements:

(1) $A(1)$ (Initial step)

(2) $(\forall n \in \mathcal{N})[A(n) \Rightarrow A(n + 1)]$ (Induction step)

That this implies $(\forall n \in \mathcal{N})A(n)$ may be reasoned as follows. By (1), $A(1)$. By (2) (as a special case) we have $A(1) \Rightarrow A(2)$. Hence $A(2)$. Again, a special case of (2) is $A(2) \Rightarrow A(3)$. Hence $A(3)$. And so on right through the natural numbers.

The thing to notice about this is that neither of the two statements we actually prove is the statement we are trying to establish. What we prove are the initial case (1) and the conditional (2). The step from these two to the conclusion $(\forall n \in \mathcal{N})A(n)$ (which I just explained) is known as the *principle of mathematical induction*.

As an example let us use the method of induction to prove a simple theorem.

THEOREM 2.5.3
If $x > 0$ then for any $n \in \mathcal{N}$,

$$(1+x)^{n+1} > 1 + (n+1)x$$

PROOF Let $A(n)$ be the statement

$$(1+x)^{n+1} > 1 + (n+1)x$$

$A(1)$ is the statement

$$(1+x)^2 > 1 + 2x$$

which is true by virtue of the binomial expansion

$$(1+x)^2 = 1 + 2x + x^2$$

together with the fact that $x > 0$.

The next step is to prove the statement

$$(\forall n \in \mathcal{N})[A(n) \Rightarrow A(n+1)]$$

To do this we take an arbitrary (!) n in \mathcal{N} and prove the conditional

$$A(n) \Rightarrow A(n+1)$$

To do this we assume $A(n)$ and try to deduce $A(n+1)$. We have

$$
\begin{aligned}
(1+x)^{n+2} &= (1+x)^{n+1}(1+x) \\
&> (1 + (n+1)x)(1+x) \quad [\text{by } A(n)] \\
&= 1 + (n+1)x + x + (n+1)x^2 \\
&= 1 + (n+2)x + (n+1)x^2 \\
&> 1 + (n+2)x \quad [\text{since } x > 0]
\end{aligned}
$$

This proves $A(n+1)$.

Hence by induction (that is to say, by the principle of mathematical induction) the theorem is proved. \square

The following summarizes the method of proof by induction. You wish to prove that some statement $A(n)$ is valid for all natural numbers n. First you establish $A(1)$. This is usually just a matter of trivial observation. You then present an algebraic argument that establishes the conditional

$$A(n) \Rightarrow A(n+1)$$

for any n. In general, you do this as follows. Assume $A(n)$. Look at the statement of $A(n+1)$, and somehow try to reduce it to $A(n)$, which has been assumed true, and thereby deduce the truth of $A(n+1)$. This accomplished, the induction proof is complete, by virtue of the principle of mathematical induction.

In setting out an induction proof formally, three points should be remembered:

(1) State clearly that the method of induction is being used.

(2) Prove the case $n = 1$ (or at the very least make the explicit observation that this case is obviously true, if this is the case).

(3) (The hard part:) Prove the conditional

$$A(n) \Rightarrow A(n+1)$$

One variant of induction that sometimes arises concerns the proof of statements such as

$$(\forall n \geq n_0)A(n)$$

where n_0 is some given natural number. In such a case, the first step of the induction proof consists not of a verification of $A(1)$ (which may not be true) but of $A(n_0)$ (the first case). The second step of the proof consists of the proof of the statement

$$(\forall n \geq n_0)[A(n) \Rightarrow A(n+1)]$$

This is what happens in the following theorem, part of the *Fundamental Theorem of Arithmetic*.

THEOREM 2.5.4
Every natural number greater than 1 is either a prime or a product of primes.

PROOF At first you might think that the statement to be proved by induction is

$$(\forall n \in \mathcal{N})A(n)$$

where

$$A(n) : n \text{ is either a prime or a product of primes}$$

However, as will shortly become clear, it is more convenient to replace $A(n)$ by the (stronger) statement

> $B(n)$: every natural number m such that $1 < m \leq n$ is either
> a prime or a product of primes

So here goes. And from now on we shall set out our proof in a mathematically correct fashion.

We prove, by induction, that $B(n)$ is true for all natural numbers $n > 1$. This clearly proves the theorem.

For $n = 2$, the result is trivial: $B(2)$ holds because 2 is prime. (Notice that in this case we must start at $n = 2$ rather than the more common $n = 1$.)

Now assume $B(n)$. We deduce $B(n + 1)$. Let m be a natural number such that $1 < m \leq n + 1$. If $m \leq n$, then by $B(n)$, m is either a prime or a product of primes. So in order to prove $B(n + 1)$ we need only show that $n + 1$ itself is either a prime or a product of primes. If $n + 1$ is a prime there is nothing further to say. Otherwise, n is composite, which means there are natural numbers p, q, such that

$$1 < p, q < n + 1$$

and

$$n + 1 = pq$$

Now $p, q \leq n$ so by $B(n)$, each of p and q is either a prime or a product of primes. But then $n + 1 = pq$ is a product of primes. This completes the proof of $B(n + 1)$.

The theorem now follows by induction. More precisely, the principle of mathematical induction yields the validity of the statement

$$(\forall n \in \mathcal{N}) B(n)$$

which trivially implies the theorem. ☐

Of course, in the above example, the conditional

$$B(n) \Rightarrow B(n + 1)$$

was rather easy to establish. (Indeed, we used $B(n)$ rather than the more obvious $A(n)$ mentioned earlier, precisely in order to carry through this simple argument.) In many cases, real ingenuity is required. But do not be misled into confusing the proof by induction of the main result

$$(\forall n \in \mathcal{N}) A(n)$$

with the technical subproof of the induction step

$$(\forall n \in \mathcal{N})[A(n) \Rightarrow A(n + 1)]$$

Without an announcement of the fact that induction is being used, and an observation or proof that $A(1)$ is valid, no amount of technical cleverness at proving the conditional $[A(n) \Rightarrow A(n + 1)]$ will amount to a proof of the statement $(\forall n \in \mathcal{N}) A(n)$.

So much for induction. One last remark concerning proofs of statements of the form

$$\forall x \, A(x)$$

If all else fails, there is the method of contradiction. By assuming $\neg\forall x \, A(x)$ we obtain an x such that $\neg A(x)$ (because $\neg\forall x \, A(x)$ is equivalent to $\exists x \neg A(x)$). Now we have a place to start. The difficulty is finding the finish (i.e., the contradiction).

Exercises 2.5

(1) Say whether each of the following is true or false, and support your decision by a *proof*:

 (a) There exist real numbers x and y such that $x + y = y$.

 (b) For all positive integers n, $n^2 + n + 41$ is a prime.

 (c) $\forall x \exists y (x + y = 0)$ (where x, y are real number variables).

 (d) $(\exists m \in \mathcal{N})(\exists n \in \mathcal{N})(3m + 5n = 12)$.

 (e) For all integers a, b, c, if a divides bc (without remainder), then either a divides b or a divides c.

 (f) The sum of any five consecutive integers is divisible by 5 (without remainder).

 (g) For any integer n, the number $n^2 + n + 1$ is odd.

 (h) Between any two distinct rational numbers there is a third rational number.

 (i) For any real numbers x, y, if x is rational and y is irrational, then $x + y$ is irrational.

 (j) For any real numbers x, y, if $x + y$ is irrational, then at least one of x, y is irrational.

 (k) For any real numbers x, y, if $x + y$ is rational, then at least one of x, y is rational.

(2) Prove that $\sqrt{3}$ is irrational.

(3) Prove that if p is a prime number then \sqrt{p} is irrational.

(4) Prove that \sqrt{n} is irrational iff n is not a perfect square.

(5) Let m and n be integers. Prove that:

 (a) If m and n are even, then $m + n$ is even.

 (b) If m and n are even, then mn is divisible by 4.

 (c) If m and n are odd, then $m + n$ is even.

(d) If one of m, n is even and the other is odd, then $m + n$ is odd.

(e) If one of m, n is even and the other is odd, then mn is even.

(6) Prove that if every even natural number greater than 2 is a sum of two primes, then every odd natural number greater than 5 is a sum of three primes. [The assumption here is known as the *Goldbach Conjecture*, and is a famous unsolved problem of long standing. Computer searches have verified the conjecture for all even numbers up to many billions.]

(7) The notation

$$\sum_{i=1}^{n} a_i$$

is a common abbreviation for the sum

$$a_1 + a_2 + a_3 + \ldots + a_n$$

For instance,

$$\sum_{r=1}^{n} r^2$$

denotes the sum

$$1^2 + 2^2 + 3^2 + \cdots + n^2$$

Prove the following by induction:

(a) $\forall n \in \mathcal{N} : \displaystyle\sum_{r=1}^{n} r = \frac{1}{2}n(n + 1)$

(b) $\forall n \in \mathcal{N} : \displaystyle\sum_{r=1}^{n} r^2 = \frac{1}{6}n(n + 1)(2n + 1)$

(c) $\forall n \in \mathcal{N} : \displaystyle\sum_{r=1}^{n} 2^n = 2^{n+1} - 2$

(d) $\forall n \in \mathcal{N} : \displaystyle\sum_{r=1}^{n} r.r! = (n + 1)! - 1$

(8) Prove the following by induction:

(a) $4^n - 1$ is divisible by 3.

(b) The sum of the first n odd numbers is equal to n^2.

(c) $(n + 1)! > 2^{n+3}$ for all $n \geq 5$.

(d) If a set A has n elements, then the power set $\mathcal{P}(A)$ has 2^n elements. (See Chapter 3 for details of set theory.)

2.6 The Integers

The integers provide a convenient domain to illustrate some of the techniques used in order to prove theorems in mathematics.

Given any two integers you can add them, subtract one from the other, or multiply them together, and the result will always be another integer. Division is not so straightforward. For some pairs of integers, say 5 and 15, division is possible (at least one way round: 15 divides by 5 to give the integer result 3). For other pairs, say 7 and 15, division is not possible unless you are prepared to allow fractional results. Purely in terms of integers, without making any recourse to non-integer rationals, what can be said is that the process of division leads to two numbers, a *quotient* and a *remainder*.

For example, if you divide 9 by 4 you get a *quotient* of 2 and a *remainder* of 1:

$$9 = 4 \cdot 2 + 1$$

This is a special case of our first formal theorem concerning integers.

For the proof, it is convenient to introduce a common notation. Given any integer a, let $|a|$ denote the number that results from dropping any minus sign. That is:

$$|a| = \begin{cases} a & \text{if } a \geq 0 \\ -a & \text{if } a < 0 \end{cases}$$

For example, $|7| = 7$ and $|-93| = 93$.

The number $|a|$ is called the *absolute value* of a. We discuss it in more depth in the next chapter.

THEOREM 2.6.1 (*The Division Theorem*)
Let a, b be integers, $b > 0$. Then there are unique integers q, r such that $a = q \cdot b + r$ and $0 \leq r < b$.

PROOF There are two things to be proved: that q, r exist with the stated properties and that such q, r are unique. We prove existence first.

The idea is to look at all non-negative integers of the form $a - kb$, where k is an integer, and show that one of them is less than b. (Any value of k for which this occurs will be a suitable q, the value r then being given by $r = a - kb$.)

Are there any such integers $a - kb \geq 0$? Yes, there are. Take $k = -|a|$. Then, since $b \geq 1$,

$$a - kb = a + |a| \cdot b \geq a + |a| \geq 0$$

Since such integers $a - kb \geq 0$ do exist, there will be a smallest one, of course. Call it r, and let q be the value of k for which it occurs, so that $r = a - qb$. To complete the (existence) proof, we show that $r < b$.

Suppose, on the contrary, that $r \geq b$. Then

$$a - (q+1)b = a - qb - b = r - b \geq 0$$

Thus $a - (q+1)b$ is a non-negative integer of the form $a - kb$. But r was chosen as the smallest one, and yet $a - (q+1)b < a - qb = r$, so this situation is contradictory. Thus it must be the case that $r < b$, as we wished to prove.

That leaves us with the proof of uniqueness of q, r. The idea is to show that if there are representations

$$a = qb + r = q'b + r'$$

of a with $0 \leq r, r' < b$, then in fact $r = r'$ and $q = q'$.

We start by rearranging the above equation as

$$(1) \quad r' - r = b \cdot (q - q')$$

Taking absolute values:

$$(2) \quad |r' - r| = b \cdot |q - q'|$$

But

$$-b < -r \leq 0 \quad \text{and} \quad 0 \leq r' < b$$

which together imply that

$$-b < r' - r < b$$

or in other words

$$|r' - r| < b$$

Hence by (2),

$$b \cdot |q - q'| < b$$

This implies that

$$|q - q'| < 1$$

There is only one possibility now, namely that $q - q' = 0$, i.e., $q = q'$. It follows at once using (1) that $r = r'$. The proof is complete. $\qquad \Box$

The above theorem applies only to the division of an integer a by a *positive* integer b. More general is:

THEOREM 2.6.2
Let a, b be integers, $b \neq 0$. Then there are unique integers q, r such that

$$a = q \cdot b + r \quad and \quad 0 \leq r < |b|$$

PROOF The case where $b > 0$ has been dealt with in Theorem 2.6.1, so assume now that $b < 0$. Since $|b| > 0$, applying Theorem 2.6.1 provides unique integers q', r' such that

$$a = q' \cdot |b| + r' \quad and \quad 0 \leq r' < |b|$$

Set $q = -q', r = r'$. Then, since $|b| = -b$, we get

$$a = q \cdot b + r \quad and \quad 0 \leq r < |b|$$

as required. □

Theorem 2.6.2 is also sometimes referred to as the Division Theorem. In each case, the number q is referred to as the *quotient* of a by b, and r is called the *remainder*.

Simple though it is, the Division Theorem yields many results that can be of assistance in computational work. For instance (and this is a very simple example), if you were faced with a search for numbers which are squares of primes, it might be helpful to know that the square of any odd number is one more than a multiple of 8. (For example, $3^2 = 9 = 8 + 1, 5^2 = 25 = 3 \cdot 8 + 1$.) To verify this fact, note that by the Division Theorem, any number can be expressed in one of the forms $4q, 4q + 1, 4q + 2, 4q + 3$, so any odd number has one of the forms $4q + 1, 4q + 3$. Squaring each of these gives

$$(4q + 1)^2 = 16q^2 + 8q + 1 = 8(2q^2 + 1) + 1$$
$$(4q + 3)^2 = 16q^2 + 24q + 9 = 8(2q^2 + 3q + 1) + 1$$

In both cases the result is one more than a multiple of 8.

In the case where division of a by b produces a remainder $r = 0$, we say that a is *divisible* by b. That is to say, an integer a is said to be *divisible* by another integer b if and only if there is an integer q such that $a = b \cdot q$. For example, 45 is divisible by 9 whereas 44 is not divisible by 9.

The theorem that follows lists the basic properties of divisibility. In stating this result we make use of the (standard) notation $b|a$ to mean that a is divisible by b.

THEOREM 2.6.3
Let a, b, c, d be integers, $a \neq 0$. Then:

(i) $a|0$, $a|a$;

(ii) $a|1$ if and only if $a = \pm 1$;

(iii) if $a|b$ and $c|d$, then $ac|bd$ (for $c \neq 0$);

(iv) *if a|b and b|c, then a|c (for b ≠ 0);*

(v) *[a|b and b|a] if and only if a = ±b;*

(vi) *if a|b and b ≠ 0, then |a| ≤ |b|;*

(vii) *if a|b and a|c, then a|(bx + cy) for any integers x, y.*

PROOF In each case the proof is simply a matter of going back to the definition of $a|b$. For instance, to prove (iv), the assumptions mean that there are integers d and e such that $b = da$ and $c = eb$, and it follows at once that $c = (de)a$, so $a|c$. To take another case, consider (vi). Since $a|b$ there is an integer d such that $b = da$. Thus $|b| = |d| \cdot |a|$. Since $b \neq 0$, we must have $d \neq 0$ here, so $|d| \geq 1$. Thus $|a| \leq |b|$, as required. The remaining cases are left as an exercise. ◻

The formal definition of a *prime* number is an integer $p > 1$ which is only divisible by 1 and p.

Exercises 2.6

(1) Prove all the parts of Theorem 2.6.3.

(2) Prove that every odd number is of one of the forms $4n + 1$ or $4n + 3$.

(3) Prove that for any integer n, at least one of the integers n, $n + 2$, $n + 4$ is divisible by 3.

(4) A classic unsolved problem in number theory asks if there are infinitely many pairs of "twin primes", pairs of primes separated by 2, such as 3 and 5, 11 and 13, or 71 and 73. Prove that the only prime triple (i.e., three primes, each 2 from the next) is 3, 5, 7.

(5) It is a standard result about primes that if p is prime, then whenever p divides a product ab, p divides at least one of a, b. Prove the converse, that any natural number having this property (for any pair a, b) must be prime.

(6) Prove that if a is an odd integer, then $24|a(a^2 - 1)$. [Hint: Look at the example that followed Theorem 2.6.2.]

(7) Prove the following version of the Division Theorem. Given integers a, b with $b \neq 0$, there are unique integers q and r such that

$$a = qb + r \quad \text{and} \quad -\tfrac{1}{2}|b| < r \leq \tfrac{1}{2}|b|$$

[Hint: Write $a = q'b + r'$ where $0 \leq r' < |b|$. If $0 \leq r' \leq \tfrac{1}{2}|b|$, let $r = r'$, $q = q'$. If $\tfrac{1}{2}|b| < r' < |b|$, let $r = r' - |b|$, and set $q = q' + 1$ if $b > 0$ and $q = q' - 1$ if $b < 0$.]

2.7 Mathematical Truth

The notion of "truth" is one that has occupied philosophers for generations, and no doubt will continue to do so, and it is not our present intention to enter into this rather bewildering area. However, we should say something about what a mathematician means by the assertion that some mathematical statement is *true* (or *false*). This requires a little knowledge of the development of mathematics.

All mathematical concepts have their origin in the everyday world around us. For example, the natural numbers arise from our desire (or need) to count collections of objects, the rational numbers are required in order to measure lengths, geometry and trigonometry assist us in navigation, and so on. At the initial stage, "true" means "experimentally verifiable".

Thus we can demonstrate the "truth" of the assertion

$$1 + 2 = 3$$

by taking one apple and two oranges and noting that there are three items altogether. But what about the assertion

$$1\,000\,000\,000 + 5 = 1\,000\,000\,005$$

Most of us would (I hope) agree that this statement is just as "true" as the previous one, though it is clearly out of the question to verify this fact by counting. So why do we so readily agree that the equation is valid?

Well, although the concepts of natural number and addition arise out of everyday experience, they are really just idealized, abstract concepts — figments of our imagination — whose properties and behavior are *postulated* (by us) to correspond to natural phenomena. (For instance, none of us has ever seen "the number 10". We may have seen ten apples in a bowl. And we have just seen two symbols on a page which we read as "ten". But "the number 10" itself remains locked away somewhere in our psyche.) The equation

$$1\,000\,000\,000 + 5 = 1\,000\,000\,005$$

is "true" because it is a consequence of the properties we postulate for the natural numbers.

Of course, addition of natural numbers is so very basic and familiar to us, that it is rare to regard matters in quite this fashion ("rare", notice, not "incorrect"). But what about (say) complex numbers? The equation

$$(1 + i)^2 = 2i$$

is "true", but we cannot test this by experiment — even a hypothetical experiment! In fact it is "true" only in the sense that it can be proved to be a consequence of the properties we postulate the complex numbers to have.

This brings us to the real nature of pure mathematics. In each area, be it the theory of addition of integers, the theory of real numbers, geometry, calculus, or whatever, we begin with a collection of *postulates* (or *axioms*) that set down the properties the objects under consideration are to have. These postulates are formulated either by observation of some everyday event (as with the postulates for addition of integers or for classical Euclidean geometry), or else by consideration of what properties the system "ought" to have. From then on "true" means simply "provable from the postulates". (There are also axioms for logical deduction, so the notion of just what constitutes a "proof" is also determined by means of postulates, but at this stage you are probably best advised to stick with whatever intuitive notions you have already regarding the nature of proof.)

In the case of the natural numbers or the rational numbers, the basic postulates are often never cited at first-year college level, so you may have some difficulty in appreciating the above remarks. But it is extremely likely that any course you take in real analysis will commence with some discussion of the postulates for the real numbers, and then you will see how "true" corresponds to "provable from the axioms". (We discuss the real numbers briefly in Chapter 5.)

My only concern now is that when you are next asked to prove something, you will be puzzled as to where you should start. Especially as you don't know "all the axioms." Don't worry! (In fact, you probably will, but what more can I say?) Except in unusual circumstances, a proof of something is simply a logical argument *that convinces us that the fact is true* (for the system concerned). The proof given earlier that $\sqrt{2}$ is irrational is valid in this sense. Admittedly this proof used various properties of the natural numbers (such as the square of an odd number being odd) that are only true because they follow from the postulates for the integers. But we all know that $odd^2 = odd$, so in this case there is no need to go right back to the axioms to check it. As a general rule: let common sense be the judge of what is or is not a proof. After a little while you should have no difficulties in deciding what you may or may not assume in any given instance. (Though even the experts can disagree on this matter.)

Chapter 3

Set Theory

3.1 Sets

The concept of a *set* is extremely basic and pervades the whole of present-day mathematical thought. Any well-defined collection of objects is a set. For instance we have:

- the set of all students in your class

- the set of all prime numbers

- the set whose members are you and my left ear

All it takes to determine a set is some way of specifying the collection.

If A is a set, then the objects in the collection A are called either the *members* of A or the *elements* of A. (With a concept as simple as a set, there is no way to avoid such circular definitions: but it should be clear what is meant.) We write

$$x \in A$$

to denote that x is an element of A.

Some sets occur frequently in mathematics, and it is convenient to adopt a standard notation for them:

\mathcal{N} : the set of all natural numbers (i.e., the numbers 1, 2, 3, etc.)

\mathcal{Z} : the set of all integers (i.e., 0 and all positive and negative whole numbers)

\mathcal{Q} : the set of all rational numbers (i.e., fractions)

\mathcal{R} : the set of all real numbers

\mathcal{R}^+ : the set of all non-negative real numbers

Thus, for example,

$$x \in \mathcal{R}$$

means that x is a real number. And

$$(x \in \mathcal{Q}) \wedge (x > 0)$$

means that x is a positive rational number.

There are several ways of specifying a set. If it has a small number of elements we can list them. In this case we denote the set by enclosing the list of the elements in curly brackets; thus, for example,

$$\{1, 2, 3, 4, 5\}$$

denotes the set consisting of the natural numbers 1, 2, 3, 4, and 5.

By use of "dots" we can extend this notation to any finite set, e.g.,

$$\{1, 2, 3, \ldots, n\}$$

denotes the set of the first n natural numbers. Again

$$\{2, 3, 5, 7, 11, 13, 17, \ldots, 53\}$$

could (given the right context) be used to denote the set of all primes up to 53.

Certain infinite sets can also be described by the use of dots (only now the dots have no end), e.g.,

$$\{2, 4, 6, 8, \ldots, 2n, \ldots\}$$

denotes the set of all even natural numbers. Again,

$$\{\ldots, -8, -6, -4, -2, 0, 2, 4, 6, 8, \ldots\}$$

denotes the set of all even integers.

In general, however, except for finite sets with only a small number of elements, sets are best described by giving the property which defines the set. If $A(x)$ is some property, the set of all those x that satisfy $A(x)$ is denoted by

$$\{x \mid A(x)\}$$

Or, if we wish to restrict the x to those which are members of a certain set X, we would write

$$\{x \in X \mid A(x)\}$$

This is read "the set of all x in X such that $A(X)$". For example:

$$\mathcal{N} = \{x \in \mathcal{Z} \mid x > 0\}$$
$$\mathcal{Q} = \{x \in \mathcal{R} \mid (\exists m, n \in \mathcal{Z})[(m > 0) \wedge (mx = n)]\}$$
$$\{\sqrt{2}, -\sqrt{2}\} = \{x \in \mathcal{R} \mid x^2 = 2\}$$
$$\{1, 2, 3\} = \{x \in \mathcal{N} \mid x < 4\}$$

Two sets, A, B are *equal*, written $A = B$, if they have exactly the same elements. As the above example shows, equality of sets does not mean they have identical definitions; there are often many different ways of describing the same set. The definition of equality reflects rather the fact that a set is just a collection of objects.

If we have to prove that the sets A and B are equal, it is often quite difficult to prove in one go that they have the same elements. What is usually done is to split the proof into two parts:

(a) Show that every member of A is a member of B.

(b) Show that every member of B is a member of A.

Taken together, (a) and (b) clearly imply $A = B$. (The proof of both (a) and (b) is usually of the "take an arbitrary element" variety. To prove (a), for instance, we must prove $(\forall x \in A)(x \in B)$; so we take an arbitrary element x of A and show that x must be an element of B.)

The set notations introduced have obvious extensions. For instance, we can write

$$Q = \{m/n \mid m, n \in \mathcal{Z}, n \neq 0\}$$

and so on.

It is convenient in mathematics to introduce a set that has no elements: the *empty set* (or *null set*). There will only be one such set, of course, since any two such will have exactly the same elements and thus be (by definition) equal. The empty set is denoted by the Scandinavian letter

$$\emptyset$$

[Note that this is not the Greek letter ϕ.] The empty set can be specified in many ways, e.g.,

$$\emptyset = \{x \in \mathcal{R} \mid x^2 < 0\}$$
$$\emptyset = \{x \in \mathcal{N} \mid 1 < x < 2\}$$
$$\emptyset = \{x \mid x \neq x\}$$

Notice that \emptyset and $\{\emptyset\}$ are quite different sets. \emptyset is the empty set: it has NO members. $\{\emptyset\}$ is a set that has ONE member. Hence

$$\emptyset \neq \{\emptyset\}$$

What is the case here is that

$$\emptyset \in \{\emptyset\}$$

(The fact that the single element of $\{\emptyset\}$ is the empty set is irrelevant in this connection: $\{\emptyset\}$ *does* have an element, \emptyset does *not*.)

A set A is called a *subset* of a set B if every element of A is a member of B. For example, $\{1, 2\}$ is a subset of $\{1, 2, 3\}$. We write

$$A \subseteq B$$

to mean that A is a subset of B. If we wish to emphasize that A and B are unequal here, we write

$$A \subset B$$

and say that A is a *proper subset* of B. (This usage compares with the ordering relations \leq and $<$ on \mathcal{R}.)

Clearly, for any sets A, B, we have

$$A = B \text{ iff } (A \subseteq B) \wedge (B \subseteq A)$$

The set of *all* subsets of a set X is called the *power set* of X, denoted by $\mathcal{P}(X)$. For example, if $X = \{1, 2, 3\}$, then

$$\mathcal{P}(X) = \{\emptyset, \{1\}, \{2\}, \{3\}, \{1, 2\}, \{1, 3\}, \{2, 3\}, \{1, 2, 3\}\}$$

(See Exercises 3.1 (6), (7) below for further examples.)

Exercises 3.1

(1) What well-known set is this:

$$\{n \in \mathcal{N} \mid (n > 1) \wedge (\forall x, y \in \mathcal{N})[(xy = n) \Rightarrow (x = 1 \vee y = 1)]\}$$

(2) Let

$$P = \{x \in \mathcal{R} \mid \sin(x) = 0\}, Q = \{n\pi \mid n \in \mathcal{Z}\}$$

What is the relationship between P and Q?

(3) Let

$$A = \{x \in \mathcal{R} \mid (x > 0) \wedge (x^2 = 3)\}$$

Give a simpler definition of the set A.

(4) Prove that for any set A:

$$\emptyset \subseteq A \text{ and } A \subseteq A$$

(5) Prove that if $A \subseteq B$ and $B \subseteq C$, then $A \subseteq C$.

(6) List all subsets of the set $\{1, 2, 3, 4\}$.

(7) List all subsets of the set $\{1, 2, 3, \{1, 2\}\}$.

(8) Can you find a set A for which $\mathcal{P}(A) = \emptyset$?

(9) Let $A = \{x \mid P(x)\}$, $B = \{x \mid Q(x)\}$, where P, Q are formulas such that $\forall x[P(x) \Rightarrow Q(x)]$. Prove that $A \subset B$.

(10) Prove that $A \subseteq B$ if and only if $\mathcal{P}(A) \subseteq \mathcal{P}(B)$.

(11) Prove (by induction) that a set with exactly n elements has 2^n subsets.

(12) Let

$$A = \{o, t, f, s, e, n\}$$

Give an alternative definition of the set A. (Hint: this is connected with \mathcal{N} but is not entirely mathematical.)

3.2 Operations on Sets

There are various natural operations we can perform on sets. (They correspond *roughly* to addition, multiplication, and subtraction for integers.)

Given two sets A, B we can form the set of all objects which are members of either one of A and B. This set is called the *union* of A and B and is denoted by

$$A \cup B$$

Formally, this set has the definition

$$A \cup B = \{x \mid (x \in A) \vee (x \in B)\}$$

(Now the reader gets the first real inkling as to why we defined our precise use of *or* to mean *inclusive-or*.) We show below how this definition can be illustrated diagrammatically.

We first of all introduce the notion of a *universal set*. It is rare in mathematics to discuss arbitrary sets. One is generally only interested in sets of reals, or sets of integers, and so on. For instance, when discussing sets of reals it can be convenient to imagine that there is nothing else in the world except reals. This prevents our having to keep saying things like "x is a real" or "A is a set of reals." When we do this, we are fixing \mathcal{R} to be the *universal set* for the duration of the discussion. This means that any set mentioned in the ensuing discussion will be *assumed* to be a subset of \mathcal{R}. Likewise if we fix \mathcal{Z} as universal set, or \mathcal{N}, etc.

Notice that there is not one universal set; rather we introduce one such whenever it is convenient to ignore all the possibilities other than the ones concerned. (If this is still confusing to you, it should become clearer when we look at set-theoretic complements in a short while.) And now to our diagram.

Fix some universal set U. We represent U by means of a rectangle on the page. The elements of U are represented by the points within the rectangle. So, in terms of this representation, relative to our fixed universal set, all objects lie inside our rectangle. Sets (i.e., subsets of U) are denoted by encircling regions within U.

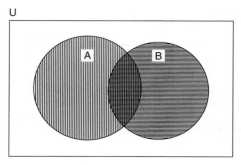

Figure 3.1: Union and intersection.

In Figure 3.1, the set A is represented by the interior of the left-hand circle, the region shaded vertically, and the set B is represented by the interior of the right-hand circle, the region shaded horizontally. The set $A \cup B$ is represented now by the total shaded region (which includes a region that is shaded two ways here). Such a diagrammatic representation of sets is called a *Venn diagram*.

The *intersection* of the sets A, B is the set of all members which A and B have in common. It is denoted by

$$A \cap B$$

and has the formal definition

$$A \cap B = \{x \mid (x \in A) \wedge (x \in B)\}$$

In Figure 3.1, the set $A \cap B$ is represented by the region which is doubly shaded.

Two sets A, B are said to be *disjoint* if they have no elements in common: that is, if $A \cap B = \emptyset$.

Once we have fixed a universal set we can introduce the notion of the *complement* of the set A. Relative to the universal set U, the complement of a set A is the set of all elements of U that are not in A. This set is denoted by A', and has the formal definition

$$A' = \{x \in U \mid x \notin A\}$$

[Notice that we write $x \notin A$ instead of $\neg(x \in A)$, for brevity.]

For instance, if the universal set is the set \mathcal{N} of natural numbers, and E is the set of even (natural) numbers, then E' is the set of odd (natural) numbers. In Figure 3.2, A' is represented by the unshaded region.

Figure 3.3 presents a Venn diagram illustrating the inclusion (subset) relation

$$A \subseteq B$$

The following theorem sums up the basic facts about the three set operations just discussed.

THEOREM 3.2.1

Let A, B, C be subsets of a universal set U.

(1) $A \cup (B \cup C) = (A \cup B) \cup C$

(2) $A \cap (B \cap C) = (A \cap B) \cap C$ *((1) and (2) are the associative laws)*

(3) $A \cup B = B \cup A$

(4) $A \cap B = B \cap A$ *((3) and (4) are the commutative laws)*

(5) $A \cup (B \cap C) = (A \cup B) \cap (A \cup C)$

(6) $A \cap (B \cup C) = (A \cap B) \cup (A \cap C)$ *((5) and (6) are the distributive laws)*

(7) $(A \cup B)' = A' \cap B'$

(8) $(A \cap B)' = A' \cup B'$ *((7) and (8) are called the De Morgan laws)*

(9) $A \cup A' = U$

(10) $A \cap A' = \emptyset$ *((9) and (10) are the complementation laws)*

(11) $(A')' = A$ *(self-inverse law)*

Figure 3.2: Complement.

Figure 3.3: Inclusion.

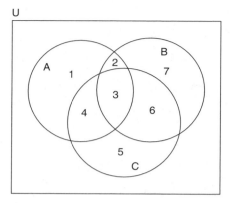

Figure 3.4: The distributive law.

PROOF The proof of parts (1) to (4) and of parts (9) to (11) are left as an exercise. For the remaining cases we give a Venn diagram proof for (5) and (7) and establish (6) and (8) using a logical argument.

Consider part (5). We can represent this diagrammatically as in Figure 3.4. We have drawn the diagram to illustrate the most general (i.e., the most complex) case, which is when all the sets have elements in common with each of the other sets. We have also numbered each of the regions. Now let us see which regions correspond to which sets.

Well,

$$A \text{ is represented by the regions } 1, 2, 3, 4$$

and

$$B \cap C \text{ is represented by the regions } 3, 6 \text{ (common to } B \text{ and } C\text{)}$$

So,

$$A \cup (B \cap C) \text{ is represented by the regions } 1, 2, 3, 4, 6$$

Again,

$$A \cup B \text{ is represented by the regions } 1, 2, 3, 4, 6, 7$$

and

$$A \cup C \text{ is represented by the regions } 1, 2, 3, 4, 5, 6$$

so

$$(A \cup B) \cap (A \cup C) \text{ is represented by the regions} $$
$$1, 2, 3, 4, 6 \text{ (common to both)}$$

But this is the same region that we obtained for $A \cup (B \cap C)$ above.

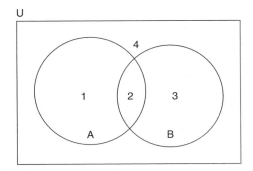

Figure 3.5: The De Morgan laws.

Consider part (6). Here is a proof using a logical argument. Let

$$D = A \cap (B \cup C)$$

$$E = (A \cap B) \cup (A \cap C)$$

We prove first that $D \subseteq E$. Let $x \in D$. Then $x \in A$ and $x \in B \cup C$. Since $x \in B \cup C$, either $x \in B$ or $x \in C$, or both. In case $x \in B$, we have $x \in A$ and $x \in B$, so $x \in A \cap B$. On the other hand, if $x \notin B$, then we must have $x \in C$, so $x \in A$ and $x \in C$, giving $x \in A \cap C$. In either of these two cases, $x \in E$. Hence $D \subseteq E$.

Now we prove $E \subseteq D$. Let $x \in E$. There are two cases. Suppose first that $x \in A \cap B$. Then $x \in A$ and $x \in B$, so $x \in A$ and $x \in B \cup C$, so $x \in D$. On the other hand, if $x \notin A \cap B$, then $x \in A \cap C$, so again we obtain $x \in A$ and $x \in B \cup C$, giving $x \in D$. Hence $E \subseteq D$.

Hence, both $D \subseteq E$ and $E \subseteq D$, and we conclude that $D = E$, proving (6).

We turn now to (7) and (8). We prove (7) by means of a Venn diagram argument. Figure 3.5 illustrates the situation.

Referring to the diagram, $A \cup B$ is represented by the regions 1, 2, 3. Thus, $(A \cup B)'$ is represented by the single remaining region 4.

Again, A' is represented by regions 3 and 4, and B' is represented by regions 1 and 4, so $A' \cap B'$ is represented by the common region 4.

This proves (7).

We prove (8) by means of a logical argument. Let $x \in (A \cap B)'$. Thus $x \notin A \cap B$. Hence

$$\neg((x \in A) \wedge (x \in B))$$

This is the same as

$$(\neg(x \in A) \vee \neg(x \in B))$$

In other words

$$(x \notin A) \vee (x \notin B)$$

i.e.,

$$(x \in A') \vee (x \in B')$$

i.e.,

$$x \in A' \cup B'$$

This shows that $(A \cap B)' \subseteq A' \cup B'$. The above steps may be reversed to give $A' \cup B' \subseteq (A \cap B)'$, thereby completing the proof. ☐

Exercises 3.2

(1) Draw a Venn diagram to illustrate the fact

$$A \cup B \subseteq C$$

(2) Draw a Venn diagram to illustrate the fact

$$A \subseteq B \cup C$$

(3) For each of the following pairs of sets A, B, find $A \cup B$ and $A \cap B$:
 (a) $\{a, b, c\}$, $\{c, d, e, f\}$
 (b) $\{x \in \mathcal{R} \mid x < 0\}$, $\{x \in \mathcal{R} \mid x \geq 0\}$
 (c) $\{1, 2, \{1, 2, 3\}\}$, $\{1, \{1, 2\}\}$

(4) Give two proofs of the fact that for any sets A, B:

$$A \cap B \subseteq A \subseteq A \cup B$$

one proof by Venn diagram, the other by a logical argument.

(5) Prove the following statements:

$$A \cap \emptyset = \emptyset$$
$$A \cup \emptyset = A$$
$$A \cap A = A \cup A = A$$

(6) Give Venn diagram proofs of Theorem 3.2.1 parts (1) to (4) and (9) to (11).

(7) Give Venn diagram proofs of Theorem 3.2.1 parts (6) and (8) and logical argument proofs of parts (5) and (7).

(8) By considering the sets $A = \mathcal{R}$, $B = \mathcal{Q}$, $C =$ the set of irrationals, show that it is not always the case that

$$(A \cup B) \cap C = A \cup (B \cap C)$$

(9) Show that the equality

$$(A \cap B) \cup C = A \cap (B \cup C)$$

is not always valid.

(10) Prove the following:

 (a) $A \subseteq B$ iff $A \cup B = B$.

 (b) $A \subseteq B$ iff $A \cap B = A$.

 (c) If $A \subseteq B \cup C$ and $A \cap B = \emptyset$, then $A \subseteq C$.

 (d) If $A \subseteq C$ and $B \subseteq C$, then $A \cup B \subseteq C$.

 (e) If $A \cup B \subseteq C \cup D$, $C \subseteq A$, and $A \cap B = \emptyset$, then $B \subseteq D$.

(11) Given two sets A, B, the *set difference* $A - B$ is defined to be the set

$$A - B = \{x \mid x \in A \wedge x \notin B\}$$

Prove the following:

 (a) $A - \emptyset = A$.

 (b) $\emptyset - A = \emptyset$.

 (c) $A - B = A \cap B'$.

 (d) $A \subseteq B$ iff $A - B = \emptyset$.

 (e) If $A \subseteq B$, then $A - C \subseteq B - C$.

(12) Prove that $\mathcal{P}(A \cap B) = \mathcal{P}(A) \cap \mathcal{P}(B)$.

(13) Prove that $\mathcal{P}(A) \cup \mathcal{P}(B) \subseteq \mathcal{P}(A \cup B)$. Give an example to show that equality is not always true here.

(14) By using the set identities established in Theorem 3.2.1, give an *algebraic* (not a set-theoretic) proof of the identity

$$[A \cup (B \cap C)] \cap C = (A \cup B) \cap C$$

(15) Use the distributive laws to give an algebraic proof of the identity

$$(A \cup B) \cap (C \cup D) = (A \cap C) \cup (A \cap D) \cup (B \cap C) \cup (B \cap D)$$

Does the result remain valid if we interchange \cap and \cup here? Justify your answer.

(16) By now, you have probably noticed some similarities between set operations, the properties of the logical connectives, and elementary integer arithmetic. The following table summarizes the situation by tabulating the various operations according to similarities in their behavior.

Arithmetic	Logic	Sets
$m + n$	$p \vee q$	$A \cup B$
$m \times n$	$p \wedge q$	$A \cap B$
$-n$	$\neg p$	A'

Carry out a systematic investigation of these comparisons with respect to the properties of commutativity, associativity, distributivity, and idempotence.

3.3 Real Intervals

Certain types of subset of the reals \mathcal{R} occur so frequently that it is convenient to introduce a special notation for them.

By an *interval* we mean an uninterrupted stretch of the real line. There are a number of different kinds of interval, for which there is a fairly widespread standard notation.

Let $a, b \in \mathcal{R}, a < b$. The *open interval* (a, b) is the set

$$(a, b) = \{x \in \mathcal{R} \mid a < x < b\}$$

The *closed interval* $[a, b]$ is the set

$$[a, b] = \{x \in \mathcal{R} \mid a \leq x \leq b\}$$

The point to notice here is that neither a nor b is an element of (a, b), but both a and b are elements of $[a, b]$. (This seemingly trivial distinction turns out to be highly significant in elementary real analysis.) Thus, (a, b) is the stretch of the real line beginning "just past" a and ending "just before" b, while $[a, b]$ is the stretch beginning with a and ending with b.

The above notation extends in an obvious manner. We call

$$[a, b) = \{x \in \mathcal{R} \mid a \leq x < b\}$$

a *left-closed, right-open* interval, and

$$(a, b] = \{x \in \mathcal{R} \mid a < x \leq b\}$$

a *left-open, right-closed* interval.

Both $[a, b)$ and $(a, b]$ are sometimes referred to as *half-open* (or *half-closed*) intervals.

Finally, we set

$$(-\infty, a) = \{x \in \mathcal{R} \mid x < a\}$$
$$(-\infty, a] = \{x \in \mathcal{R} \mid x \leq a\}$$
$$(a, \infty) = \{x \in \mathcal{R} \mid x > a\}$$
$$[a, \infty) = \{x \in \mathcal{R} \mid x \geq a\}$$

Notice that the symbol ∞ is never coupled with a square bracket. This would be misleading, since ∞ is *not* a number, just a useful symbol. In the above definitions it simply helps us to extend a convenient notation to cover another case.

Exercises 3.3

(1) Prove that the intersection of two intervals is again an interval. Is the same true for unions?

(2) Taking \mathcal{R} as the universal set, express the following as simply as possible in terms of intervals and unions of intervals:

(a) $[1, 3]'$ (b) $(1, 7]'$
(c) $(5, 8]'$ (d) $(3, 7) \cup [6, 8]$
(e) $(-\infty, 3)' \cup (6, \infty)$ (f) $\{\pi\}'$
(g) $(1, 4] \cap [4, 10]$ (h) $(1, 2) \cap [2, 3)$
(i) A', where $A = (6, 8) \cap (7, 9]$ (j) A', where $A = (-\infty, 5] \cup (7, \infty)$

3.4 Absolute Values

The *absolute value* of a real number x is defined by:

$$|x| = \begin{cases} x & \text{if } x \geq 0 \\ -x & \text{if } x < 0 \end{cases}$$

In geometric terms, $|x|$ is the distance of the point x from the origin (on the real line). For later use, we prove some simple results about this concept.

THEOREM 3.4.1
For any a,

$$|a| = \sqrt{a^2}$$

PROOF Clearly, $|a|^2 = a^2$. Taking positive square roots, $\sqrt{|a|^2} = \sqrt{a^2}$. But $|a| \geq 0$. Hence $\sqrt{|a|^2} = |a|$ and the theorem is proved. ⬜

THEOREM 3.4.2
For any a, b,

$$|ab| = |a||b|$$

PROOF There are four possible cases:

(1) $a \geq 0, b \geq 0$. Then $|ab| = ab = |a||b|$.

(2) $a \geq 0, b < 0$. Then $|ab| = -(ab) = a(-b) = |a||b|$.

(3) $a < 0, b \geq 0$. Then $|ab| = -(ab) = (-a)b = |a||b|$.

(4) $a < 0, b < 0$. Then $|ab| = ab = (-a)(-b) = |a||b|$. ⬜

THEOREM 3.4.3
Let $a > 0$. Then for any x,

$$|x| \leq a \quad iff \ -a \leq x \leq a$$

PROOF Exercise for the reader. ⬜

THEOREM 3.4.4 (Triangle inequality)
For any a, b,

$$|a + b| \leq |a| + |b|$$

PROOF Clearly, by Theorem 3.4.2,

$$ab \leq |ab| = |a||b|$$

Thus

$$2ab \leq 2|a||b|$$

So

$$a^2 + 2ab + b^2 \leq a^2 + 2|a||b| + b^2$$

So by Theorem 3.4.1,

$$(a + b)^2 \leq |a|^2 + 2|a||b| + |b|^2$$

i.e.,

$$(a + b)^2 \leq (|a| + |b|)^2$$

Hence

$$\sqrt{(a + b)^2} \leq \sqrt{(|a| + |b|)^2}$$

So, by Theorem 3.4.1,

$$|a + b| \leq |a| + |b|$$

and the proof is complete. ◻

If $x, y \in \mathcal{R}$, the positive number $|x - y|$ represents the distance between x and y on the real line. If $x \neq y$, then $|x - y| > 0$. Conversely, if $|x - y| > 0$, then $x \neq y$. Hence $x = y$ iff $|x - y| = 0$. By the triangle inequality, if $x, y, z \in \mathcal{R}$, we have

$$|x - y| \leq |x - z| + |z - y|$$

This inequality is also referred to as *the triangle inequality*.

3.5 Inequalities

You can solve an inequality in much the same way you can solve an equation. Whereas the solution to an equation is generally one or more real numbers, the solution to an inequality is typically one or more intervals.

You can add the same quantity to both sides of an inequality without affecting the validity of the inequality. Likewise, you can subtract the same quantity from both sides. For example, if

$$x \leq y$$

then, for any number a,

$$x + a \leq y + a \quad \text{and} \quad x - a \leq y - a$$

If you multiply both sides of a valid inequality by a positive quantity, it remains valid. If you multiply both sides of a valid inequality by a negative quantity, the inequality reverses direction. For example, if

$$x < y$$

then, for any positive number a,

$$ax < ay \quad \text{but} \quad (-a)x > (-a)y$$

Since division by a non-zero number a is the same as multiplication by $1/a$, the last remark also applies to division.

An inequality of the form

$$|E(x)| \leq M$$

(where $E(x)$ is some expression involving x) that involves an absolute value can be replaced by the pair of inequalities

$$-M \leq E(x) \leq M$$

that does not involve an absolute value.

When you are faced with solving an inequality, you will often find it useful to split it into two or more cases, depending on the signs of the various constituent expressions.

Worked Examples

(1) Solve the inequality

$$x^2 + 3x - 4 > 0$$

SOLUTION Factor the quadratic, to give ·

$$(x - 1)(x + 4) > 0$$

There are two cases to consider.

Case 1. $x - 1 > 0$ and $x + 4 > 0$.
 That is, $x > 1$ and $x > -4$. In this case, both inequalities are satisfied if $x > 1$.

Case 2. $x - 1 < 0$ and $x + 4 < 0$.
 That is, $x < 1$ and $x < -4$. In this case, both inequalities are satisfied if $x < -4$.
 Thus, the complete solution is the pair of intervals $(-\infty, -4)$, $(1, \infty)$.

(2) Solve the inequality

$$\frac{3x + 4}{x - 3} < 2$$

SOLUTION The idea is to split it into cases.

Case 1. $x > 3$. Since $x - 3 > 0$, the inequality becomes

$$3x + 4 < 2x - 6$$

which simplifies to $x < -10$. This is incompatible with the assumption for Case 1 (namely $x > 3$), so there is no solution in this case.

Case 2. $x < 3$. The inequality becomes

$$3x + 4 > 2x - 6$$

which simplifies to $x > -10$. Now, in Case 2 we also have the assumption $x < 3$. So the overall solution set is the interval

$$(-10, 3)$$

(3) Solve the inequality

$$\frac{3x - 4}{x + 2} < 5$$

SOLUTION The idea is to split it into two cases, depending on the sign of the denominator: $x + 2 < 0$ and $x + 2 > 0$.

Case 1. $x > -2$.
 Multiplying through by the denominator, which is positive, the inequality becomes

$$3x - 4 < 5(x + 2)$$

which simplifies to

$$3x - 4 < 5x + 10$$

Hence, $-14 < 2x$, giving the solution $x > -7$. This is automatically satisfied in this case, since $x > -2$. Thus the solution interval in this case is $(-2, \infty)$.

Case 2. $x < -2$.
 Multiplying through by the denominator, which is negative, the inequality becomes

$$3x - 4 > 5(x + 2)$$

which simplifies to

$$3x - 4 > 5x + 10$$

Hence, $-14 > 2x$, giving the solution: $x < -7$. Thus, in this case, the solution interval is $(-\infty, -7)$.
 Thus the entire solution to the inequality is

$$(-\infty, -7), (-2, \infty)$$

(4) Solve the inequality

$$\left| \frac{1}{x} - 2 \right| < 5$$

SOLUTION The idea is to rewrite the inequality as

$$-5 < \frac{1}{x} - 2 < 5$$

Adding 2 to each term, this becomes

$$-3 < \frac{1}{x} < 7$$

Now split into two cases depending on the sign of x.

Case 1. $x > 0$.
 Multiplying through by x, the inequality becomes

$$-3x < 1 < 7x$$

Since $x > 0$ in this case, $-3x < 1$ is automatically true.
Thus the inequality reduces to $x > \frac{1}{7}$.

Case 2. $x < 0$.
 Multiplying through by x, the inequality becomes

$$-3x > 1 > 7x$$

The inequality $1 > 7x$ is automatically true in this case, since $x < 0$.
Thus the inequality reduces to $x < -\frac{1}{3}$.
Thus the entire solution set is

$$\left(-\infty, -\frac{1}{3}\right), \left(\frac{1}{7}, \infty\right)$$

Exercises 3.5

(1) Solve the following inequalities:
 (a) $|5x - 1| \geq 2$
 (b) $x^2 > x + 2$
 (c) $(x + 1)(x + 2)(x + 3) > 0$
 (d) $\dfrac{x}{x + 1} < 1$
 (e) $\left|\dfrac{3x + 2}{x + 3}\right| > 3$

(2) Using a graphing calculator, graph each of the functions in examples 1, 2, and
 3 in the above Examples section, and see how the behavior of the graph relates
 to the solution to the inequality.

3.6 Arbitrary Unions and Intersections

The associative laws allow us to form unions and intersections of arbitrary finite collections of sets by iterating the operations of union and intersection of pairs of sets. For example:

$$A_1 \cup A_2 \cup \ldots \cup A_{n-1} \cup A_n = (\ldots((A_1 \cup A_2) \cup \ldots) \cup A_{n-1}) \cup A_n$$

However, in mathematics we often need to consider unions and intersections of *infinite* collections of sets, and in this case the above approach will not work. Instead, we have to go back to the original ideas behind set union and intersection and formulate an alternative definition.

There are two distinct notations that are commonly used, depending on the context. Suppose first of all that to each element i of some set I there corresponds a set A_i. Then we refer to the collection

$$\{A_i \mid i \in I\}$$

as an *indexed family* (or *indexed set*) of sets and the set I as the *index set*.

For example, if P is the set of all prime numbers and

$$E_i = \{n \in \mathcal{N} \mid i \text{ is a factor of } n\}$$

for each $i \in P$, then we get the indexed family

$$\{E_i \mid i \in P\}$$

Again, if for each real number r, X_r denotes the set of all rational numbers less than r, then we get the indexed family

$$\{X_r \mid r \in \mathcal{R}\}$$

Notice that in the first example, the family can be thought of as *enumerated* as an infinite sequence A_1, A_2, A_3, \ldots (by taking $A_1 = E_2$, $A_2 = E_3$, $A_3 = E_5$, etc.), but this is not possible for the second example.

The *union* of an indexed family $\{A_i \mid i \in I\}$ is defined to be the set of all those elements that lie in one or more of the sets A_i in the family. It is generally denoted by

$$\bigcup_{i \in I} A_i$$

Thus:

$$\bigcup_{i \in I} A_i = \{a \mid (\exists i \in I)[a \in A_i]\}$$

The *intersection* of an indexed family $\{A_i \mid i \in I\}$ is defined to be the set of all those elements that lie in all of the sets A_i in the family, and is generally denoted by

$$\bigcap_{i \in I} A_i$$

Thus:

$$\bigcap_{i \in I} A_i = \{a \mid (\forall i \in I)[a \in A_i]\}$$

It should be noticed that in the case where the index set I consists of just two elements, say $I = \{1, 2\}$, then:

$$\bigcup_{i \in I} A_i = A_1 \cup A_2 \quad \text{and} \quad \bigcap_{i \in I} A_i = A_1 \cap A_2$$

so the notions of arbitrary union and intersection over indexed families are generalizations of the notions of union and intersection of pairs of sets (and hence also of finite unions and intersections obtained by iteration of pair union and pair intersection).

Although the following theorem is essentially trivial, its proof provides a good illustration of the manner in which arbitrary unions and intersections are handled.

THEOREM 3.6.1
Let $\{A_i \mid i \in I\}$ be an indexed family of sets. Then for any $i_0 \in I$

$$A_{i_0} \subseteq \bigcup_{i \in I} A_I \quad \text{and} \quad \bigcap_{i \in I} A_I \subseteq A_{i_0}$$

PROOF Suppose $x \in A_{i_0}$. Then $(\exists i \in I)[x \in A_i]$ (in particular, i_0 is such an i). So by definition, $x \in \bigcup_{i \in I} A_I$. This proves the first inclusion.

Now suppose $x \in \bigcap_{i \in I} A_I$. Then $(\forall i \in I)[x \in A_i]$. So in particular, $x \in A_{i_0}$. This proves the second inclusion. ◻

Notice that the above proofs made no mention of whether the index set I was finite or infinite. Nor was any mention made of any *order* on the sets A_i in terms of the first set, the second set, etc. Indeed, the index set I might not have any particular ordering, so there may be no natural way to order the indexed family. The argument depended solely upon the definitions of indexed union and intersection in terms of quantifiers over the index set. The same is true for the proofs involved in the following theorem:

THEOREM 3.6.2
Let $\{A_i \mid i \in I\}$ be an indexed family of sets, and let B be any set, all subsets of some universal set U. Then:

(1) $\displaystyle B \cup \bigcup_{i \in I} A_i = \bigcup_{i \in I} (B \cup A_i)$

(2) $B \cap \bigcap_{i \in I} A_i = \bigcap_{i \in I} (B \cap A_i)$

(3) $B \cap \bigcup_{i \in I} A_i = \bigcup_{i \in I} (B \cap A_i)$

(4) $B \cup \bigcap_{i \in I} A_i = \bigcap_{i \in I} (B \cup A_i)$

(5) $\left(\bigcup_{i \in I} A_i \right)' = \bigcap_{i \in I} A_i'$

(6) $\left(\bigcap_{i \in I} A_i \right)' = \bigcup_{i \in I} A_i'$

PROOF We prove (1), (3), and (5) and leave the others as an exercise.

(1) Let $x \in B \cup \bigcup_{i \in I} A_i$. If $x \in B$, then $x \in B \cup A_i$ for all $i \in I$, so $x \in \bigcup_{i \in I} (B \cup A_i)$. Otherwise, $x \in \bigcup_{i \in I} A_i$, so for some $i_0 \in I$, $x \in A_{i_0}$. Then $x \in B \cup A_{i_0}$, so again $x \in \bigcup_{i \in I} (B \cup A_i)$.

Conversely if $x \in \bigcup_{i \in I} (B \cup A_i)$, then for some $i_0 \in I$, $x \in B \cup A_{i_0}$. If $x \in B$, then $x \in B \cup \bigcup_{i \in I} A_i$. Otherwise, $x \in A_{i_0}$, so $x \in \bigcup_{i \in I} A_i$, and again $x \in B \cup \bigcup_{i \in I} A_i$.

(3) Let $x \in B \cap \bigcup_{i \in I} A_i$. Then $x \in B$ and for some $i_0 \in I$, $x \in A_{i_0}$. Thus $x \in B \cap A_{i_0}$, and it follows that $x \in \bigcup_{i \in I} (B \cap A_i)$.

Conversely, suppose $x \in \bigcup_{i \in I} (B \cap A_i)$. Then for some $i_0 \in I$, $x \in B \cap A_{i_0}$. Hence $x \in B$ and $x \in A_{i_0}$. Thus $x \in B$ and $x \in \bigcup_{i \in I} A_i$. Thus $x \in B \cap \bigcup_{i \in I} A_i$.

(5) Let $x \in U$ be arbitrary. Then

$$x \in \left(\bigcup_{i \in I} A_i \right)' \quad \text{iff } x \notin \bigcup_{i \in I} A_i$$
$$\text{iff } \neg (\exists i \in I)[x \in A_i]$$
$$\text{iff } (\forall i \in I)[x \notin A_i]$$
$$\text{iff } (\forall i \in I)[x \in A_i']$$
$$\text{iff } x \in \bigcap_{i \in I} A_i' \qquad \qquad \Box$$

In the case where the index set is the set \mathcal{N} of natural numbers, we often write

$$\bigcup_{i=1}^{\infty} A_i \quad \text{and} \quad \bigcap_{i=1}^{\infty} A_i$$

in place of $\bigcup_{i \in \mathcal{N}} A_i$ and $\bigcap_{i \in \mathcal{N}} A_i$, respectively.

An alternative notation for arbitrary unions and intersections is used when the family of sets concerned is not indexed. Suppose that \mathcal{F} is some collection of sets. Then we write

$$\bigcup \mathcal{F}$$

for the set

$$\{x \mid (\exists A \in \mathcal{F})[x \in A]\}$$

and

$$\bigcap \mathcal{F}$$

for the set

$$\{x \mid (\forall A \in \mathcal{F})[x \in A]\}$$

Thus, in the case where $\mathcal{F} = \{A_i \mid i \in I\}$, we have:

$$\bigcup \mathcal{F} = \bigcup_{i \in I} A_i$$

$$\bigcap \mathcal{F} = \bigcap_{i \in I} A_i$$

Exercises 3.6

(1) Complete the proof of Theorem 3.6.2.

(2) For each $n \in \mathcal{N}$, let $A_n = \{n\}$. What are $\bigcup_{n \in \mathcal{N}} A_n$ and $\bigcap_{n \in \mathcal{N}} A_n$?

(3) For each $r \in \mathcal{R}^+$, let $A_r = \{x \in \mathcal{R} \mid 0 \leq x < r\}$, where \mathcal{R}^+ denotes the set of all strictly positive reals. What are $\bigcup_{r \in \mathcal{R}^+} A_r$ and $\bigcap_{r \in \mathcal{R}^+} A_r$?

(4) Let \mathcal{F} be a family of sets. Prove that $\bigcup \mathcal{F}$ is the smallest set X such that $A \subseteq X$ for all $A \in \mathcal{F}$, in the sense that:

- $A \subseteq \bigcup \mathcal{F}$ for all $A \in \mathcal{F}$;
- if $(\forall A \in \mathcal{F})[A \subseteq X]$, for any set X, then $\bigcup \mathcal{F} \subseteq X$.

(5) Formulate and prove a similar characterization of $\bigcap \mathcal{F}$.

3.7 Cartesian Products

The related notions of *ordered pairs* and the *Cartesian product* of two sets generalizes the idea of the two-dimensional Euclidean plane and the Cartesian coordinate system in plane geometry.

The *ordered pair* of two objects a and b (presented in that order) is an object (a, b) associated with these objects (in that order) such that two ordered pairs (a, b) and (x, y) are *equal* if and only if $a = x$ and $b = y$.

The use of the adjective "ordered" distinguishes the ordered pair (a, b) of a, b from the *unordered pair* (usually just referred to as the *pair*) $\{a, b\}$. When we write the pair set $\{a, b\}$, the order in which the elements a, b are presented is not important: $\{a, b\} = \{b, a\}$. But for the ordered pair, we have $(a, b) = (b, a)$ only in the trivial case $a = b$; if $a \neq b$, then $(a, b) \neq (b, a)$.

You will doubtless have noticed that the notation adopted for ordered pairs is the same as that used to denote open intervals of the real line. This is an unfortunate clash, since the notions of open intervals and ordered pairs are quite different. Of course, in the case where the objects a, b are not real numbers, the expression (a, b) is not ambiguous; it must mean the ordered pair. But if a, b are reals, care has to be exercised in interpreting this expression. Usually the context is enough to make the intended meaning clear.

Given two sets A and B, the *Cartesian product* of A, B is defined to be the set:

$$A \times B = \{(a, b) \mid a \in A \wedge b \in B\}$$

For example, if $A = \{1, 2\}$ and $B = \{2, 3, 4\}$, then

$$A \times B = \{(1, 2), (1, 3), (1, 4), (2, 2), (2, 3), (2, 4)\}$$

Notice that Cartesian product is in general not commutative. In the case of the above example, we have:

$$B \times A = \{(2, 1), (2, 2), (3, 1), (3, 2), (4, 1), (4, 2)\}$$

So, for instance, we have $(1, 2) \in A \times B$ and $(1, 2) \notin B \times A$.

THEOREM 3.7.1
Let A, B, C, D be sets. Then:

(a) $A \times (B \cup C) = (A \times B) \cup (A \times C)$

(b) $A \times (B \cap C) = (A \times B) \cap (A \times C)$

(c) $(A \times B) \cap (C \times D) = (A \cap C) \times (B \cap D)$

(d) $(A \times B) \cup (C \times D) \subseteq (A \cup C) \times (B \cup D)$

PROOF (a) We present a "double-inclusion proof". Suppose first that $(x, y) \in A \times (B \cup C)$. Then $x \in A$ and $y \in B \cup C$. If $y \in B$, then $(x, y) \in A \times B$; if $y \in C$, then $(x, y) \in A \times C$. In either case, $(x, y) \in (A \times B) \cup (A \times C)$. Thus $A \times (B \cup C) \subseteq (A \times B) \cup (A \times C)$.

Conversely, suppose $(x, y) \in (A \times B) \cup (A \times C)$. If $(x, y) \in A \times B$, then $x \in A$ and $y \in B$, so $y \in B \cup C$ and hence $(x, y) \in A \times (B \cup C)$; if $(x, y) \in A \times C$, then $x \in A$ and $y \in B \cup C$, so again $(x, y) \in A \times (B \cup C)$. Thus $(A \times B) \cup (A \times C) \subseteq A \times (B \cup C)$.

This proves (a).

(b) We present a single "iff" proof, rather than the "double-inclusion" proof used for part (a). For any ordered pair (x, y) (of elements of the appropriate universal set), we have:

$$(x, y) \in A \times (B \cap C) \quad \text{iff } x \in A \wedge y \in (B \cap C)$$
$$\text{iff } x \in A \wedge [y \in B \wedge y \in C]$$
$$\text{iff } [x \in A \wedge y \in B] \wedge [x \in A \wedge y \in C]$$
$$\text{iff } [(x, y) \in A \times B] \wedge [(x, y) \in A \times C]$$
$$\text{iff } (x, y) \in (A \times B) \cap (A \times C)$$

(c) Let (x, y) be any ordered pair (of elements of the appropriate universal set). Then:

$$(x, y) \in (A \times B) \cap (C \times D) \quad \text{iff } (x, y) \in A \times B \text{ and } (x, y) \in C \times D$$
$$\text{iff } x \in A \text{ and } y \in B \text{ and } x \in C \text{ and } y \in D$$
$$\text{iff } x \in A \cap C \text{ and } y \in B \cap D$$
$$\text{iff } (x, y) \in (A \cap C) \times (B \cap D)$$

(d) Notice that this is an inclusion, not an identity. (The proof that these two sets are not always equal is left as one of the assigned exercises.) Let $(x, y) \in (A \times B) \cup (C \times D)$. Then $(x, y) \in A \times B$ or $(x, y) \in C \times D$. In the former case, we have $x \in A$ and $y \in B$, so certainly $x \in A \cup C$ and $y \in B \cup D$, giving $(x, y) \in (A \cup C) \times (B \cup D)$, and similarly in the other case. This proves (d). ⬚

The notion of an ordered pair generalizes to what is known as an (ordered) *n-tuple*. The *n*-tuple of n objects a_1, a_2, \ldots, a_n (given in that order) is an object (a_1, a_2, \ldots, a_n) having the property that

$$(a_1, a_2, \ldots, a_n) = (b_1, b_2, \ldots, b_n) \text{ iff } [a_1 = b_1 \wedge a_2 = b_2 \wedge \ldots \wedge a_n = b_n]$$

For example, $(1, 2, 3, 4, 5)$, $(5, 4, 3, 2, 1)$, $(5, 4, 3, 1, 2)$ are all different 5-tuples. A 3-tuple is sometimes called a *triple* (or *ordered triple*).

Exercises 3.7

(1) Let $A = \{1, 3, 5\}$, $B = \{2, 4\}$. Find $A \times A$, $B \times B$, $A \times B$, $B \times A$.

(2) Show that the inclusion

$$(A \times B) \cup (C \times D) \subseteq (A \cup C) \times (B \cup D)$$

proved in Theorem 3.7.1 cannot be extended to an identity.

(3) Let A, B be non-empty sets. Prove that $A \times B = B \times A$ iff $A = B$.

(4) Is the above result true if one of A, B is empty? Explain.

(5) Show that if A, B, C are non-empty, then $A \times (B \times C) \neq (A \times B) \times C$.

(6) Prove that if A has m elements and B has n elements, then $A \times B$ has mn elements.

3.8 The Historical Development of Set Theory

Unlike most areas of mathematics, which arise as a result of the cumulative efforts of many mathematicians, sometimes over several generations, set theory is the creation of a single individual, George Cantor, around the turn of the 20th century. Like many before him, Cantor was troubled by the concept of infinity in mathematics.

Such concerns go back at least as far as the ancient Greeks. Around 540 B.C. Zeno of Elea formulated a number of paradoxes to illustrate the problems associated with infinity. The best known of these is the paradox of Achilles and the tortoise. Achilles is to race the tortoise over a 100 meter track. Since Achilles can run 10 times as fast as the tortoise, he gives his slower rival a 10 meter start. The question is, how far has Achilles run when he draws level with the tortoise, as surely he will. Common sense tells us the answer will be close to 11 meters. Zeno clearly knew this, but he presented an argument to show that Achilles never catches up with the tortoise, and challenges us to explain why this argument is not valid. Here is that argument.

At the start, the tortoise is 10 meters in front. When Achilles has covered the 10 meters that originally separates them, the tortoise has advanced 1 meter. He remains ahead, this time by 1 meter. By the time Achilles has covered that 1 meter, the tortoise has advanced 1/10 meter. He remains ahead by 1/10 meter. By the time Achilles has covered that 1/10 meter, the tortoise has advanced another 1/100 meter. He remains ahead. Clearly, we can continue to reason in this way *ad infinitum*, demonstrating that the tortoise always remains ahead of Achilles.

Modern mathematics resolves this puzzle by having the ability to conceive of the number

$$10 + 1 + 1/10 + 1/100 + 1/1000 + \ldots$$

where the addition continues forever. As a decimal, we represent the answer to this infinite sum as

$$11.111\ldots$$

where the dots indicate that the decimal expansion continues forever. This is the exact distance Achilles has covered when he draws level with the tortoise. But the idea of a number represented by an infinitely long decimal was not available to the ancient Greeks. Such ideas began to emerge only in the 19th century.

To this day, the notion of an infinite decimal is a difficult one to come to grips with. Many students, when asked to say whether the number

$$0.999\ldots$$

is either equal to 1 or else "just less than 1", choose the latter. That answer is not wrong. It depends on how the question is interpreted formally within mathematics. If the number system you are working with is the real numbers, which is the normal assumption in mathematics, then $0.999\ldots$ is exactly equal to 1. However, there is an alternative approach that formalizes measurement in terms of what are called the *hyper-reals*, and in that framework, $0.999\ldots$ does indeed turn out to be "just less than 1". The point is — and this is what came out of Cantor's groundbreaking work — the notion of infinity is so problematic that you have to resort to a formally defined treatment, based on axioms, and different choices of axioms give rise to different results.

Since this book is aimed at preparing you for further study in traditional modern mathematics, where the real numbers are taken as basic, the "right" answer you should remember is that $0.999\ldots$ is equal to 1. Here is a quick proof of that fact. Let

$$x = 0.999\ldots$$

Multiplying by 10,

$$10x = 9.999\ldots$$

Subtract the first equation from the second, to give:

$$9x = 9$$

Now divide by 9 to conclude that $x = 1$, as claimed.

This proof depends on the fact that the multiplication and subtraction operations work on infinite decimals the way they do for finite decimals. In the real number system they do, but that fact itself requires proof. In this respect, the proof I have just given you is incomplete. But it's all you are getting from me on this subject right now, since we need to get back to the historical stuff.

In 1847, the Czech mathematician Bernhard Bolzano considered the idea of set, which he described this way:

> ... an embodiment of the idea or concept which we conceive when we regard
> the arrangement of its parts as a matter of indifference.

Bolzano allowed for the possibility of infinite sets, observing that, unlike finite sets, the elements of an infinite set could be put in one-to-one correspondence with elements of one of its proper subsets. For example, you can put the natural numbers into one-to-one correspondence with the even natural numbers (a proper subset of the former) by the simple process of associating each natural number n with its double, $2n$. Such a correspondence is never possible with a finite set. This fact was used to provide a definition of infinity: an infinite set is one whose elements can be put into

one-one correspondence with a proper subset. A finite set, then, is any set that is not infinite according to this definition.

As we saw in Chapter 1, around the middle of the 19th century, mathematicians also began to use the concept of an abstract set in order to define powerful new algebraic concepts such as groups, rings, and fields. Until Cantor came onto the scene, however, no one carried out a systematic investigation of the notion of a set itself; there was no *theory of sets*. All that changed with Cantor.

Cantor's early work was in number theory; he published a number of articles on this topic between 1867 and 1871. Then, in 1872, on a trip to Switzerland, he met Richard Dedekind, one of the Göttingen revolutionaries we encountered in Chapter 1. Dedekind was grappling with infinite sets in order to try to develop a theory of the real numbers that could explain why the methods of calculus worked.

Although Cantor and Dedekind do not appear to have collaborated on any mathematical research, over the years 1873–1879 the two exchanged a number of letters addressing deep philosophical issues about the nature of mathematical abstraction.

In 1874, Cantor published an article in a famous mathematical journal, *Crelle's Journal*, which constituted the birth of set theory. In that paper, he showed that there are at least two different kinds of infinity. Prior to this revolutionary claim, all infinite collections were assumed to be the same size, whatever that meant. Specifically, Cantor proved that the real numbers cannot be put into one-one correspondence with the natural numbers (and thus the real numbers constitutes a bigger category of infinite set than the natural numbers). The argument he gave in his paper used nested intervals. In a later paper, published in 1891, he gave a different, and much simpler proof, which is the one generally presented to mathematics students today. That simpler proof uses the method of contradiction to show that there are "more" real numbers between 0 and 1 than there are natural numbers. You will meet that proof in Chapter 4 (Theorem 1). One advantage of the alternative proof was that it could be easily generalized to show that there is an unending hierarchy of larger and larger infinities — infinitely many different infinities, in fact.

Cantor's new ideas were controversial from the very start. Leopold Kronecker, a leading (and highly influential) Prussian mathematician at the time and one of the editors of *Crelle's Journal*, was unhappy about the revolutionary new theory. Kronecker, it should be pointed out, went on record as insisting that irrational numbers do not exist. Today such a view seems absurd, but at the time the issue was still very much in debate, with mathematicians undecided as to exactly which way the subject should go when it came to dealing with the infinite. Cantor's suggestion that there was a hierarchy of different infinities was anathema to Kronecker's more finitist way of thinking.

When Cantor submitted a second paper on set theory in 1878, opposition had grown so vehement that he considered withdrawing the paper, but when Karl Weierstrass expressed his support, he allowed it to go forward and the paper was finally accepted. He did not however send any further work to that particular journal for publication.

In his 1878 paper, Cantor introduced the idea of two sets being *equivalent*, or "having the same power", meaning that they can be put into one-one correspondence.

He proved that the rational numbers have the smallest infinite power and that $\Re^n = \Re \times \cdots \times \Re$ has the same power as \Re. He also proved that if A_1, A_2, \ldots are copies of \Re, then $\bigcup_{i \in \mathcal{N}} A_i$ has the same power as \Re. You will see mathematical details of some of these ideas and results in Chapter 4.

Cantor published a six-part treatise on set theory, which appeared in the journal *Mathematische Annalen* between 1879 and 1884. In this series of papers, he discussed well-ordered sets, ordinal numbers (defined as the order types of well-ordered sets), and an early treatment of multiplication and addition of transfinite numbers. On the controversial philosophical issues, he stressed that in mathematics any concepts may be introduced, defined in terms of previously accepted concepts, provided only that they are free of contradiction.

Cantor introduced the convention of using the Hebrew letter \aleph (*aleph*) to denote infinite cardinal numbers, with \aleph_0 being the smallest infinite cardinal (the cardinality of the set of natural numbers), \aleph_1 the next, \aleph_2 the third one, and so on, *ad infinitum*!

An obvious question Cantor was unable to answer was: Which one of these infinite cardinal numbers was the cardinality of \Re, the set of real numbers? In particular, was it \aleph_1, the first uncountable cardinal? It was not until 1963 that this question was finally resolved. The American mathematician Paul Cohen proved that the question could not be answered. That's right, Cohen gave a rigorous mathematical proof that showed that the statement *"The cardinality of \Re is \aleph_1"* is neither true nor false (on the basis of what by then were the generally accepted axioms for set theory).

In 1884, Cantor attempted to move to the University of Berlin, which would offer him the prestige he felt his work deserved, but this attempt was blocked by Kronecker and others. This opposition to his career advancement led to a nervous breakdown. In that one year alone, he sent the influential Swedish mathematician Magnus Gösta Mittag-Leffler no fewer than 52 letters attacking Kronecker. Although he recovered from the crisis, his subsequent work never reached the heights of his remarkable 10-year period from 1874 to 1884.

In 1895 and 1897, Cantor published his final double treatise on set theory. The introduction looks very much like a modern set theory textbook, except that he did not use the modern symbol \in to mean "is an element of". That notation was introduced by Guiseppe Peano in 1889. It comes from the first letter of the Greek word meaning "is".

The year 1897 saw the appearance of the first of several paradoxes that arise within set theory once infinite sets are allowed. That first paradox, discovered by the mathematician Cesare Burali-Forti, has to do with the notion of well-ordered sets, and I won't go into details here, but it did mark the beginning of a series of discoveries that eventually led to the formulation of axioms for set theory.

Cantor himself discovered a paradox in 1899, this one to do with cardinalities (or powers) of sets.

The most devastating paradox was discovered by Bertrand Russell in 1902. Russell considered the set

$$R = \{X \mid X \text{ is not a member of } X\}$$

He then asked: Is R an element of R? The answers yes and no both lead to a contradiction. The question cannot be answered. The very notion of set construction itself appears to give rise to paradox.

Russell wrote to his colleague Gottlob Frege telling him about the paradox. Frege was in the process of completing his magnum opus on the foundations of arithmetic, the culmination of his life's work. Russell's paradox destroyed the bulk of that work. Frege continued with publication, but added an acknowledgement that read:

> A scientist can hardly meet with anything more undesirable than to have the foundation give way just as the work is finished. In this position I was put by a letter from Mr. Bertrand Russell as the work was nearly through the press.

By this stage, however, set theory was beginning to have a major impact on other areas of mathematics. To give just one example, in 1901, Henri Lebesgue used ideas from set theory in his definition of an abstract *measure*, and a year later used that notion, together with other set theoretic ideas, to define his *Lebesgue integral*. The train had already left the station. With so much powerful — and useful — new mathematics being heavily dependent on Cantor's set theory, there was no going back. Instead of dismissing set theory because of the paradoxes, as Kronecker wished, an urgent search began to find ways of keeping the main features of set theory and yet avoiding the paradoxes.

In 1908, Ernst Zermelo produced what looked like an acceptable solution: a set of axioms for set theory that appeared to avoid the paradoxes. A short while later, Abraham Fraenkel proposed an amendment to Zermelo's theory, and the mathematical community was finally able to breathe a collective sigh of relief. Zermelo–Fraenkel set theory became the accepted axiomatic formulation of Cantor's ideas, and remains so to this day.

Although the Zermelo–Fraenkel axioms do lead to a theory of sets that is adequate for the needs of modern mathematics, the price to be paid for stepping into the realm of infinite collections has been brought home to us many times since Cohen's 1963 breakthrough result, by the discovery of a seemingly unending sequence of seemingly simple questions about infinite sets (and mathematics built upon the set concept) that are provably unanswerable on the basis of the axioms. These are known collectively as the "undecidability results in set theory" or more generally "undecidability results in mathematics."

Exercises 3.8

(1) The Hilbert Hotel, named after the famous German mathematician (and set theory pioneer) David Hilbert, has an infinite number of rooms. The rooms are numbered 1, 2, 3, etc. One evening — it was a Super Bowl weekend — all the rooms were booked. But late that night, an unexpected VIP turned up and asked for a room. Fortunately, the desk clerk had been reading Keith Devlin's

book *Sets, Functions, and Logic*. With a bit of rearranging, the clerk was able to arrange matters so that the VIP got a room and so did everyone else. How did he do it?

(2) The following evening the rooms were all still booked, but an infinite number of new guests arrived, all without reservations. The clever desk clerk was again able to arrange matters so that all the new arrivals got rooms and so too did the guests who had advance reservations. What did the clerk do this second night?

(3) Let

$$R = \{x \mid x \notin x\}$$

Prove that the assumption that R is a well-defined set leads to a contradiction.

(4) The axioms of Zermelo–Fraenkel set theory prohibit any set from being a member of itself. But there are alternative set theories that allow sets to be members of themselves. Give an example of a collection of objects in everyday life that is a member of itself.

Chapter 4

Functions

4.1 The Function Concept

The concept of a function from real numbers to real numbers will almost certainly be familiar to you already. For instance, the equation

$$y = x^2 + x + 1$$

determines a function from real numbers to real numbers. We can draw its graph as in Figure 4.1. We are able to plot the graph of this function because, for each value of x in \mathcal{R}, we can calculate the corresponding value of y: we just substitute the value of x concerned into the equation and carry out the computation. What the above equation does is provide us with a *general rule* for calculating values.

Functions from real numbers to real numbers defined by equations in this fashion are special cases of the general *function* concept that we introduce next.

Let A and B be any non-empty sets. A *function from A to B* is a rule that associates with each member of A a unique member of B.

We make no restrictions on the rule; the only crucial points are that:

- the rule must associate a member of B to *every* element of A;

- the element of B that we associate with a given member a of A must be *unique* for that a.

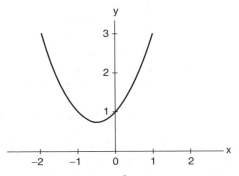

Figure 4.1: Graph of the function $y = x^2 + x + 1$.

Figure 4.2: The three parts of a function.

We write

$$f : A \rightarrow B$$

to indicate that f is a function from A to B.

Now, in the case of a function as above, given by an equation, we can refer to that function by first of all stipulating that it is *from* real numbers *to* real numbers, and secondly giving its equation. But our definition allows for all kinds of rules, not just equations. We give a number of examples presently. But first it is convenient to introduce some further nomenclature and a little notation.

If $a \in A$, we denote the unique element of B that the rule associates to a by

$$f(a)$$

We refer to the element a of A here as an *argument* of the function, and the corresponding element $f(a)$ of B as the *value* of the function at that argument (or sometimes the *image* of the point a under f).

Thus, in our original example above, we have $f : \mathcal{R} \rightarrow \mathcal{R}$ and

$$f(x) = x^2 + x + 1$$

and if a is any argument for f, then the *value* of the function f *at* that argument a is the real number $a^2 + a + 1$. For instance, the value $f(5)$ associated with argument 5 is 31, and the value $f(-2)$ associated with argument -2 is 3.

Notice that we have not defined the concept of *a function*, but rather *a function from a set A to a set B*. This distinction should be borne in mind at all times, as it affects our subsequent definitions to a very great degree.

You can think of a function as a sort of "black box" (Figure 4.2) that has an input slot and an output slot. At the input slot there is a tray, A, full of possible inputs (the arguments), and at the output slot there is a second tray, B, designed to take all the possible outputs (the values). We may feed into the black box any of the members of A. The black box then processes our input and produces a single output, which will always be a member of the set B.

It should be stressed that whereas it is the rule f that actually produces the value for a given argument, both the input tray A and the output tray B are integral parts of the function. (You cannot purchase a black box without these two attachments!) The importance of this stipulation will become apparent in due course.

If $f : A \rightarrow B$, we refer to the set A as the *domain* of f and the set B as the *codomain*.

4.2 Examples of Functions

We give a number of examples of functions. Notice the different ways of specifying the rule involved.

(1) First of all, there are the functions $f : \mathcal{R} \to \mathcal{R}$ defined by polynomials, such as the example $f(x) = x^2 + x + 1$ given above.

(2) Then there are the rational functions $f : \mathcal{R} \to \mathcal{R}$ such as

$$f(x) = \frac{x^3 - 4x^2 + x - 11}{(x^2 + 1)^5}$$

(3) The *absolute value* function $f : \mathcal{R} \to \mathcal{R}^+$ is defined by

$$f(x) = \begin{cases} x & \text{if } x \geq 0 \\ -x & \text{if } x < 0 \end{cases}$$

(where $\mathcal{R}^+ = \{x \in \mathcal{R} \mid x \geq 0\}$). This function is not defined by an equation: rather the rule consists of two separate cases. In words, the rule is: given x as argument, the value is x if $x \geq 0$ and is $-x$ if $x < 0$. Since for each x there is exactly one possible value for $f(x)$ (because no x satisfies both $x \geq 0$ and $x < 0$), this does indeed define a function. We usually write $|x|$ instead of $f(x)$, and call $|x|$ the *absolute value* of x (or sometimes the *modulus* or *mod*) of x.

(4) There are the so-called analytic functions from \mathcal{R} to \mathcal{R} such as $f(x) = \sin x$, $f(x) = \log |x|$, $f(x) = e^x$.

(5) You can combine various kinds of function $f : \mathcal{R} \to \mathcal{R}$ to form hybrids such as

$$f(x) = \frac{\sin^2 4x - 3x^2 + 21}{\cos 2x + e^{2x}} + \log |x|$$

(6) Consider next the function $f : \mathcal{R} \to \{0, 1\}$ defined by

$$f(x) = \begin{cases} 1 & \text{if } x \text{ is irrational} \\ 0 & \text{if } x \text{ is rational} \end{cases}$$

Try to draw a graph of this function.

(7) A function that occurs both in number theory and in computer science is the *greatest integer function* $f : \mathcal{R} \to \mathcal{Z}$ defined by:

$$f(x) = \text{the largest integer } z \text{ such that } z \leq x$$

Draw a graph of this function.

(8) Given any two sets A, B and any element $b_0 \in B$, there is a *constant function* $f : A \to B$ defined by:

$$f(a) = b_0 \quad (\forall a \in A)$$

(9) For any set A, the *identity function*, I_A, on A is defined by:

$$I_A(a) = a \quad (\forall a \in A)$$

(10) Let $A = \{a, b, c, d\}$, $B = \{c, d, e\}$. Define $f : A \to B$ by

$$f(a) = c, \; f(b) = d, \; f(c) = c, \; f(d) = e$$

(In this case the rule is determined by listing all the argument and value pairs individually.)

(11) If U is a universal set and A is any subset of U, then associated with A is the *characteristic function* of A, $\chi_A : U \to \{0, 1\}$, defined by:

$$\chi_A(x) = \begin{cases} 1 & \text{if } x \in A \\ 0 & \text{if } x \notin A \end{cases}$$

(12) Let A, B be non-empty sets. The *canonical projections*, π_A, π_B, from the Cartesian product set $A \times B$ onto the sets A, B, respectively, are defined by

$$\pi_A(a, b) = a \quad (\forall (a, b) \in A \times B)$$
$$\pi_B(a, b) = b \quad (\forall (a, b) \in A \times B)$$

(13) If U is a universal set and V is any subset of U, then we can define a function $f : \mathcal{P}(U) \to \mathcal{P}(V)$ by:

$$f(A) = A \cap V \quad (\forall A \subseteq U)$$

(14) Let U be the set of all countries in the world and C the set of all cities, and define $f : U \to C$ by:

$$f(a) = \text{ the capital city of } a \quad (\forall a \in U)$$

(15) Let P denote the following computer program:

```
begin
var A:integer;
read A;
print (A*A + A + 1);
end
```

Then P specifies a function $f : \mathcal{Z} \to \mathcal{Z}$. (Actually, this is only correct in theory. In practice, the computer system on which the program is run will have a maximum integer input/output size (often known as MAXINT), and then the function determined by the program *running on that machine* has domain and codomain $\{n \in \mathcal{Z} \mid -\text{MAXINT} \leq n \leq \text{MAXINT}\}$.)

(16) To take another example from computer science, let C be a compiler for a programming language L and a computer K. Let P be the set of all properly written programs in L. Let F be the set of all machine language programs for the computer K. Then C determines a function from P to F. For each source-code program p in P, the compiler produces an object-code, machine language program $C(p)$ that can be run on K.

(17) Let T be the set of all books in a library, and consider the following instruction:

Given any book b, count all the commas that occur in b.

This determines a function $f : T \to \mathcal{N}$. The rule involved in this function can also be expressed as:

$$f(b) = \text{ the number of commas that occur in } b$$

(18) Let P be the set of all people, and define $f : P \to P$ by:

$$f(p) = \text{ the father of } p \quad (\forall p \in P)$$

(19) One final example. Suppose T is a telephone directory, and let A be the set of all listed subscribers. Assume every telephone number consists of 7 digits, and let N be the set of all 7-digit numbers. Consider the following procedure. Given the name and address of a particular subscriber, look up his or her telephone number on the directory. Does this determine a function f from A to N? The answer is "maybe." There are two things that could go wrong.

First of all, a subscriber might have an unlisted number. The above rule would not produce a result for such inputs. This violates a basic requirement for a function that every input from the domain leads to an output value.

Suppose now we eliminate this problem by redefining the set A to consist of only those subscribers with listed numbers. Now do we have a function?

Again, it is possible that we do not. Some subscribers might have two or more numbers listed. This violates the requirement for a function that any input argument results in only one output value.

We may eliminate this second problem by modifying the procedure so that for each listed subscriber, look up the *first* telephone number listed for that subscriber. Now the procedure does determine a function from A to N.

We end this section with a brief discussion about what it means for two functions to be equal.

Consider the following two definitions of functions.

- f has domain \mathcal{N}, codomain \mathcal{N}, and rule

$$f(n) = n^2 - 2n + 1$$

- g has domain \mathcal{N}, codomain \mathcal{N}, and rule

$$g(n) = (n - 1)^2$$

The functions f and g have the same domain and the same codomain, but different rules. The first rule involves a squaring, a doubling, a subtraction, and an addition; the second consists of a subtraction and a squaring. If you were to follow each rule (mentally or using a hand calculator) to evaluate, say, $f(3)$ and $g(3)$, you would perform different sequences of operations.

But of course, as everyone knows, the two functions f, g will give rise to exactly the same values in every case. That is to say, for all $n \in \mathcal{N}$, $f(n) = g(n)$. Should we then consider f and g as different functions?

The answer is that we do *not*, in fact, regard these two functions as different. The notion of *equality* that we apply to functions in mathematics renders the two functions f and g as being *equal*.

In general, we say two functions f, g are *equal* iff they have the same domain (say A), the same codomain (say B), and for every $a \in A$, $f(a) = g(a)$.

Thus, equality of functions depends upon the domains being equal, the codomains being equal, and the actions being equal. We do *not* require that the rules are the same.

This definition of function equality means that we should not really speak of a function as *being* a rule that takes arguments from the domain and produces values in the codomain. Rather a function is *determined* by such a rule. It is not the rule itself that is the function, even assuming we are careful to specify the domain and codomain (as we should). It is the argument-to-value *association* the rule determines that is "the function."

Exercises 4.2

(1) Find ten further examples of functions, five mathematical, five connected with everyday life. In each case, be sure to specify the domain, the codomain, and the rule, and verify that your example is indeed a function.

(2) Give four functions, two mathematical two non-mathematical, specifying each by two different rules.

(3) Find, if possible, examples of functions $f : \mathcal{N} \to \mathcal{N}$ having the following properties:

(a) $m < n \to f(m) < f(n)$

(b) $m < n \to f(m) > f(n)$

(c) $m < n \to f(m) = f(n)$

(d) if m, n are odd, then $m < n \to f(m) < f(n)$ but if m, n are even, then $f(m) = f(n)$. What happens (for your example) in the case where one of m, n is even and the other odd?

(4) Most hand calculators provide a square root function $\sqrt{\ }$ for real numbers x. Give a precise definition of this function.

(5) Define $f : \mathcal{R} \to \mathcal{R}$ and $g : \mathcal{R} \to \mathcal{R}$ by

$$f(x) = \text{maximum}(x, -x)$$
$$g(x) = |x|$$

Prove that f and g are equal.

(6) Explain why the following functions f and g are not equal:

$$f(x) = x - 2, \, g(x) = (x^2 - 4)/(x + 2)$$

(The variable x ranges over real numbers.)

(7) Given two sets A, B with $A \subseteq B$, the *inclusion function* $i : A \to B$ is defined by

$$i(a) = a \quad (\forall a \in A)$$

What is the relationship between this function and the identity function I_A on A? Be as precise as you can.

(8) Let the universe be \mathcal{R} and let $A = \{x \in \mathcal{R} \mid 1 \le x < 5\}$. Sketch the graph of each of the following characteristic functions:

(a) χ_A (b) $\chi_{A'}$ (c) $\chi_{\{2\}}$ (d) $\chi_{\mathcal{N}}$

(9) Let U be the universe, and let $A \subseteq U$ be such that $A \ne \emptyset$ and $A \ne U$. Identify each of the following sets.

(a) $\{x \in U \mid \chi_A(x) = 1\}$
(b) $\{x \in U \mid \chi_A(x) = 0\}$
(c) $\{x \in U \mid \chi_A(x) = -1\}$

4.3 History of the Modern Function Concept

A complete history of the modern concept of a function would take some time and would have to include reference to a great many mathematicians whose work contributed in one way or another. But arguably the two most significant figures in the story are Jean Baptise Joseph Fourier, who lived in France from 1768 to 1830, and Johann Peter Gustav Lejeune Dirichlet, a Belgian mathematician who lived from 1805 to 1859.

Fourier was a good looking, dashing figure, an idealist and a powerful orator, who became active in the French Revolution. As a revolutionary, he was sufficiently independent minded that he expressed horror at the later bloody excesses, which led

to his arrest and imprisonment, and brought him perilously close to being himself guillotined. In his later years he was a friend of Napoleon.

As a young pupil at school, he showed great ability in mathematics, but when he was 19 he decided to abandon the subject and train for the priesthood. Two years later he returned to mathematics, and progressed rapidly. In 1797, he was appointed to a chair in mathematics at the new (and soon to be prestigious) Ecole Polytechnique in Paris. A year later he joined Napoleon's army as scientific adviser, traveling to Cairo as part of the French invasion of Egypt, and subsequently becoming a leading member of the French administration there, handling scientific and educational affairs.

Fourier returned to Paris in 1799 to resume his work in mathematics, but soon after Napoleon appointed him Prefect in Genoble. While there, however, he found time to continue his mathematical work, in particular carrying out an extensive study of heat flow. Although this work is now rightly recognized as a great mathematical advance, and Fourier's methods nowadays provide an indispensable tool in many areas of mathematics, physics, and engineering, at the time it caused great controversy. What was at issue was the way Fourier handled functions, expanding them as infinite trigonometric series — what are nowadays called Fourier series. Fourier's radical new methods pushed the boundaries of what mathematicians regarded as a function.

Among a number of mathematicians inspired by Fourier's work to reflect on the nature of functions, Dirichlet was the one who made the final decisive move that gave us the modern function concept.

Dirichlet made his early reputation by solving the case $n = 5$ of Fermat's Last Theorem, and went on to make a number of major contributions to algebra. His groundbreaking work on the application of calculus-based methods to the study of the natural numbers led to him being regarded by many mathematicians today as the father of the important branch of mathematics known as analytic number theory.

It was in 1837 that Dirichlet proposed that mathematicians should stop thinking of functions as computational processes, but instead concentrate simply on their overall behavior. He wrote:

> If a variable y is so related to a variable x that whenever a numerical value is assigned to x, there is a rule according to which a unique value of y is determined, then y is said to be a function of the independent variable x.

Although couched in terms of real-valued functions of a real variable, this is clearly the modern definition of a function. It was very much in line with the new approach to mathematics propogated by the Göttingen school, which we met in Chapter 1. Dirichlet was offered Gauss' chair in mathematics at Göttingen following the death of the great German mathematician in 1855. By all accounts Dirichlet was a gifted teacher. Riemann was one of his students.

Because of his work on Fourier's ideas, Dirichlet is generally considered the founder of the theory of Fourier series. Such was the novelty of Fourier's ideas that it took several leading mathematicians, culminating with Dirichlet, to put the work onto a sound footing. Riemann's *Habilitationschrift* (a post-doctoral dissertation

required to become a professor in a German university) was on the topic of Fourier series, and in the introduction he wrote that it was Dirichlet

> ... *who wrote the first profound paper about the subject.*

For today's beginning student of modern mathematics, the main lesson to be learned from all of this is that the high degree of abstraction that you have to master, of which the general definition of a function is just one example, did not come about easily, nor without cause. When you think that Fourier's work — which forced mathematicians to reconceptualize the notion of a function — lies at the heart of modern music synthesizers, Internet transfers of music files, MP3 music players, and a host of other applications, including the original questions of heat flow that motivated Fourier himself, you realize that the abstraction was the result of some decidedly practical applications.

You would do well to bear this in mind as you progress through this book. Most students I have taught this material to become frustrated at what they at first see as "meaningless game-playing with definitions." They are eager to progress through this stage and "get to the real math." Well, there is no doubt that the "new fangled abstract stuff" you will find in this book *is* playing games with definitions. But it is far from meaningless. The abstractness and the broad generality of the notions that mathematicians developed in the 19th century have proved to be extremely powerful. Master these few key ideas and you will find that you have at your disposal some very powerful machinery of wide applicability.

4.4 One-One and Onto Functions

One of the requirements that any function must fulfill (in order to *be* a function) is that any argument produces at most one value (in fact, exactly one value). Although advanced-level mathematics textbooks sometimes speak of "multiple-valued functions," such terminology is restricted to highly specialized purposes, and may be ignored here. For our purposes (and for practically all of mathematics), no function can produce two different values for the same argument. If it does, it simply is not a function.

This requirement should not be confused with the possibility of different arguments producing the same value. This can, and does, happen quite often. That is to say, for a function $f : A \to B$, it may happen that two different arguments a_1, a_2 both produce the same value, i.e., $f(a_1) = f(a_2)$.

For example, the function $f : \mathcal{R} \to \mathcal{R}$ defined by $f(x) = x^2$ is a well-defined function. In particular, for every real number *argument* there is exactly one *value*. But *different arguments* can lead to *the same value*: for instance, $f(1) = f(-1) = 1$.

If a function $f : A \to B$ is such that it never happens that different arguments lead to the same value, we say that f is *one-one*, or *injective* (or that f is an *injection*).

Thus, f is one-one if distinct inputs produce distinct outputs. In symbols:
$f : A \to B$ is one-one iff

$$(\forall a, b \in A)[a \neq b \to f(a) \neq f(b)]$$

Alternatively, we may express this condition using the contrapositive:
$f : A \to B$ is one-one iff

$$(\forall a, b \in A)[f(a) = f(b) \to a = b]$$

Notice that whether or not a function is one-one can depend critically on the domain of that function (and not just on the rule). The function $f : \mathcal{R} \to \mathcal{R}$ defined by $f(x) = x^2$ is not one-one (as observed above), but the function $g : \mathcal{R}^+ \to \mathcal{R}$ defined by $g(x) = x^2$ *is* one-one: any two distinct non-negative real arguments do produce distinct values. The rule used to define f and g is the same, but they differ in their domains.

Now let's turn our attention to the codomain, the set B of "possible" values (or outputs) for a function $f : A \to B$. I enclosed the word "possible" in quotation marks just now for the following reason. The codomain is an integral part of the specification of a function (the output tray is an integral part of our black-box apparatus). But it is possible that not every element of B is actually produced by the function (some of the space in the output tray B is not needed, the black box will never be able to fill B). If every member of B is in fact the value of the function at some argument (that is, if it *is* possible to fill tray B), we say f is *onto*, or *surjective* (or that f is a *surjection*).

In symbols, a function $f : A \to B$ is *onto* iff

$$(\forall b \in B)(\exists a \in A)[f(a) = b]$$

Note the order of the quantifiers in the above condition. For every b in B it must be possible to find an a in A such that $f(a) = b$.

For example, the function $f : \mathcal{R} \to \mathcal{R}$ defined by $f(x) = x^2$ is not onto, since no negative real is the value of the function at any argument: the function only produces non-negative values. On the other hand, the function $g : \mathcal{R} \to \mathcal{R}^+$ defined by the same formula $g(x) = x^2$ is onto, since every non-negative real is the value of the function at some argument. (What argument?)

A *pre-image* of an element $b \in B$ for a function $f : A \to B$ is any a element of A for which $f(a) = b$. For example, if $f : \mathcal{R} \to \mathcal{R}$ is defined by $f(x) = x^2 + 3$, then 1 and -1 are both pre-images of 4.

Clearly, a function $f : A \to B$ is one-one if each element of B has *at most one* pre-image under f. And f is onto if *every* element of B has a pre-image under f.

We illustrate the above notions by means of our original example of the rule determined by the equation

$$y = x^2 + x + 1$$

This determines a function $f : \mathcal{R} \to \mathcal{R}$ defined by $f(x) = x^2 + x + 1$.

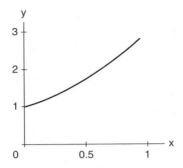

Figure 4.3: Graph of the function $y = x^2 + x + 1$ for $x \geq 0$.

Is f one-one? That is, can we find distinct reals x_1 and x_2 such that $f(x_1) = f(x_2)$ (in which case it will not be one-one), or is it the case that

$$(\forall x_1, x_2 \in \mathcal{R})[x_1 \neq x_2 \rightarrow f(x_1) \neq f(x_2)]$$

(in which case it will be one-one). A quick glance at Figure 4.1 shows us that

$$f(-1) = f(0) = 1$$

Hence $f : \mathcal{R} \rightarrow \mathcal{R}$ is not one-one.

Is f onto? That is, does every element of \mathcal{R} occur as a value of f for some argument? Again, Figure 4.1 provides the answer. The number 0 is never a value of f. Indeed, no number less than 0.75 is a value of f. Hence $f : \mathcal{R} \rightarrow \mathcal{R}$ is not onto.

The same equation $y = x^2 + x + 1$ does however determine a function $g : \mathcal{R}^+ \rightarrow \mathcal{R}$. The only difference between this function and the previous one is that our new function has a different domain, namely the non-negative reals. (We are no longer allowed to input negative numbers into our "black box.") The graph of this function is illustrated in Figure 4.3.

The function $g : \mathcal{R}^+ \rightarrow \mathcal{R}$ so defined is clearly one-one. It is indeed increasing: if $x_1 < x_2$, then $g(x_1) < g(x_2)$. But it is still not onto, for 0 is not a value, and neither is any real number less than 1.

Finally, consider the function $h : \mathcal{R}^+ \rightarrow A$ defined by the same formula $h(x) = x^2 + x + 1$, but with the codomain

$$A = \{x \in \mathcal{R} \mid x \geq 1\}$$

This function is (clearly) both one-one and onto.

A function that is one-one and onto is sometimes called a *bijection*.

Let us now briefly review our definitions. A *function* from a non-empty set A to a set B is (determined by) a rule that associates with each element a of A a unique element $f(a)$ of B, called the *value* of f at a. We write $f : A \rightarrow B$ to indicate that f is a function from A to B. The set A is the *domain* of f; the set B is the *codomain*. The elements of A are called *arguments* of the function. In specifying a function, it is not enough to give the rule: the domain and codomain must also be given.

The function $f : A \to B$ is *one-one* (or *injective*) if distinct arguments in A give rise to distinct values in B. And f is *onto* (or *surjective*) if every element of B is a value of f for some argument. A function which is both one-one and onto is called a *bijection*.

Functions are also called *mappings, maps,* or *transformations* (although each of these words tends to be reserved for special kinds of functions).

Let $f : A \to B$. If $X \subseteq A$, we call the set

$$\{f(x) \mid x \in X\}$$

the *range* of f *on* X. Thus the range of f on a subset X of the domain is the set of all values of f for arguments taken from the set X. An alternative way to express this definition is as the set

$$\{b \in B \mid (\exists x \in X)(f(x) = b)\}$$

The range of f on X is denoted by

$$f[X]$$

We refer to the set $f[A]$ simply as *the range* of f.

Clearly, an alternative definition of an *onto* function is one for which the range and the codomain coincide.

Exercises 4.4

(1) Let $A = \{1, 2\}$, $B = \{1, 2, 3\}$. List all the functions from A to B.

(2) Identify those functions in question (1) which are

 (a) one-one

 (b) onto

 (c) bijective

 In each case, identify the range of the function.

(3) For each of the functions listed in Section 4.2, say whether the function is one-one, onto, or both, and identify the range. Justify your answers.

(4) Let $f : \mathcal{R} \to \mathcal{R}$ be defined by $f(x) = x^2 - 1$.

 (a) What is the value of f at -10?

 (b) What is $f(-10)$?

 (c) What argument of f is associated with value 80?

 (d) What is $f(80)$?

 (e) What is the image of 4 under f?

 (f) What is a pre-image of 15?

 (g) List all pre-images of 12.

 (h) List all pre-images of -9.

(5) Define $f : \mathcal{R} \to \mathcal{R}$ by

$$f(x) = \frac{1}{2}(x + |x|)$$

Evaluate $f(0)$, $f(1)$, $f(2)$. Is f one-one? Is f onto? Justify your answers. What is the range of f on (a) \mathcal{R}; (b) \mathcal{R}^+?

(6) Which of the following functions are one-one?

 (a) $f : \mathcal{R} \to \mathcal{R}$, $f(x) = 5x + 6$

 (b) $f : \mathcal{R} \to \mathcal{R}$, $f(x) = x^3$

 (c) $f : \mathcal{R} \to \mathcal{R} \times \mathcal{R}$, $f(x) = (x, x)$

 (d) $f : \mathcal{R} \to \mathcal{R}$, $f(x) = \sqrt{x}$

 (e) $f : \mathcal{Z} \times \mathcal{Z} \to \mathcal{Z}$, $f(m, n) = m - n$

 Which of these functions is onto?

(7) Define $f : \mathcal{R} \to \mathcal{R}$ by

$$f(x) = \begin{cases} x^2 + 1 & \text{if } x \geq 0 \\ x - 1 & \text{if } x < 0 \end{cases}$$

Prove that f is one-one. Is f onto? What is the range of f on (a) \mathcal{R}; (b) \mathcal{R}^+?

(8) Define $f : \mathcal{N} \to \mathcal{N}$ by

$$f(n) = \begin{cases} 2n & \text{if } n \text{ is even} \\ n & \text{if } n \text{ is odd} \end{cases}$$

Show that f is one-one. Is f onto?

(9) Let $A = \{1, 3, 5, 7, \ldots\}$, the set of odd natural numbers, and $B = \{2, 4, 6, 8, \ldots\}$, the set of even natural numbers. Give examples of functions from A to B which are:

 (a) One-one but not onto.

 (b) Onto but not one-one.

 (c) Neither one-one nor onto.

 (d) One-one and onto.

(10) For each of your examples in the previous question, identify the range of the function on (a) $\{1, 3, 5, 7\}$; (b) A.

(11) Define $f : \mathcal{N} \to \mathcal{N}$ by

$$f(n) = \begin{cases} 2 & \text{if } n \text{ is not prime} \\ n & \text{if } n \text{ is prime} \end{cases}$$

Let P be the set of all primes, K the set of all non-primes (in \mathcal{N}). Identify the sets $f[\mathcal{N}]$, $f[P]$, $f[K]$.

(12) Let $f : A \to B$. Show that f is not one-one iff there are elements $x, y \in A$ such that $x \neq y$ and $f[\{x, y\}]$ has only one element.

(13) Define $f : \mathcal{R} \to \mathcal{R}$ by $f(x) = x^2 + x + 1$. What is $f[\mathcal{R}]$? What is $f[\mathcal{R}^+]$?

(14) Let $f : \mathcal{N} \to \mathcal{N}$. Show that there is a set $A \subseteq \mathcal{N}$ such that $h : A \to \mathcal{N}$ is one-one, where we define

$$h(n) = f(n) \quad (\forall n \in A)$$

Show further that there is a set $B \subseteq \mathcal{N}$ such that $g : A \to B$ is a bijection, where we define

$$g(n) = h(n) \quad (\forall n \in A)$$

4.5 Composition and Inverse Functions

Suppose $f : A \to B$ and $g : B \to C$ are given. Then we can define a function $h : A \to C$ by the following rule: given $a \in A$, use rule f to calculate $b = f(a)$ and then use rule g to calculate $g(b)$ for this b; the resulting element of C (being uniquely defined) shall be $h(a)$. In symbols,

$$h(a) = g(f(a)) \quad (\forall a \in A)$$

We denote the function h so defined by $g \circ f$ and refer to this function as the *composition* of g and f.

More generally, we can define the composition $g \circ f$ from A to C whenever $f : A \to B$ and $g : E \to C$ where E is a set such that $f[A] \subseteq E$. The important point is that $g(b)$ should be defined for every value b of f.

For example, if $f : \mathcal{R} \to \mathcal{R}$ is defined by $f(x) = 2x + 1$ and $g : \mathcal{R} \to \mathcal{R}$ is defined by $g(x) = x^2 + \sin x$, then $g \circ f : \mathcal{R} \to \mathcal{R}$ is defined, and can be computed by means of the formula

$$g \circ f(x) = (2x + 1)^2 + \sin(2x + 1)$$

It should be noted that the order in which the two functions are written in the composition $g \circ f$ is the reverse of their order of application, which is f followed by g in this composition. The notation is chosen to accord with the normal notation for function application, which reads:

$$g \circ f(x) = g(f(x))$$

An important question connected with functions is whether it is possible to "undo" the changes produced by a function, that is to say, to go backwards from the value to the argument that produced it. The formal notion is that of invertibility.

Let $f : A \to B$. We say f is *invertible* if there exists a function $g : B \to A$ such that for all $a \in A$ and all $b \in B$

$$f(a) = b \Leftrightarrow g(b) = a$$

In which case we call such a function g an *inverse* of f.

For example, in the case of the function $f : \mathcal{R} \to \mathcal{R}$ defined by $f(x) = 2x + 3$, there is an inverse function, namely $g : \mathcal{R} \to \mathcal{R}$ defined by $g(x) = (x - 3)/2$.

In the case of a function given in the form $y = f(x)$, finding an inverse amounts to "solving" for x. Thus, in the example above, if we express the original function f in the equational form

$$y = 2x + 3$$

then solving for x gives

$$x = (y - 3)/2$$

In other words, the inverse function is the function $g : \mathcal{R} \to \mathcal{R}$ defined by $g(y) = (y - 3)/2$. The only difference between this formulation of the definition of g and the first one is the use of x as the variable the first time, y the second time. In specifying a function, one is of course free to use any variable whatsoever, since the variable is just a kind of place marker, where particular arguments can be substituted in order to calculate the value. Since it is common to write real functions using x as variable, we first of all wrote the definition of g in terms of x.

Another way of defining the notion of an inverse is that g is an *inverse* of a function $f : A \to B$ iff $g : B \to A$ and

$$g \circ f = I_A \quad \text{and} \quad f \circ g = I_B$$

where I_A, I_B are the identity functions on A, B, respectively. (The proof of this equivalent definition is left as an exercise.)

THEOREM 4.5.1
Let $f : A \to B$. Then f is invertible iff it is a bijection. Moreover, if f is invertible, its inverse function is unique.

PROOF Suppose f is invertible. We prove that f is a bijection. By the invertibility of f, there is an inverse function $g : B \to A$. To show f is one-one, let $a_1, a_2 \in A$, $a_1 \neq a_2$. We prove that $f(a_1) \neq f(a_2)$. Suppose, on the contrary, that

$$f(a_1) = f(a_2) = b$$

Since g is an inverse of f, we have $g(b) = a_1$ and $g(b) = a_2$. But g is a function. Hence $a_1 = a_2$, a contradiction. Thus $f(a_1) \neq f(a_2)$, and f is shown to be one-one. To show f is onto, let $b \in B$. We show that $f(a) = b$ for some $a \in A$. Let $a = g(b)$. By the properties of g, $f(a) = f(g(b)) = b$, so we are done. Hence f is a bijection.

Conversely, assume now that f is a bijection. We show that f is invertible. Let $b \in B$. Since f is onto, there is an $a \in A$ such that $f(a) = b$. But f is one-one, so there cannot be another such a. Hence, for each $b \in B$ there is exactly one $a \in A$ with $f(a) = b$. So we can define a function $g : B \to A$ by the rule: for each $b \in B$, let $g(b)$ be the unique element a of A for which $f(a) = b$. Clearly, g is an inverse of f.

The above proof also establishes the uniqueness of the inverse, by demonstrating that for each $b \in B$ there is exactly one possible $a \in A$ to take for $g(b)$. ▯

The unique inverse of a bijective function f is usually denoted by f^{-1}.

Exercises 4.5

(1) Let $f : A \to B$, $g : B \to C$, $h : C \to D$. Prove that

$$h \circ (g \circ f) = (h \circ g) \circ f$$

(Associative law)

(2) Define $f : \mathcal{R} \to \mathcal{R}$ by

$$f(x) = \begin{cases} x^2 & \text{if } x \geq 0 \\ x - 1 & \text{if } x < 0 \end{cases}$$

Define $g : \mathcal{R} \to \mathcal{R}$ by

$$g(x) = \begin{cases} x + 1 & \text{if } x \geq 1 \\ 2x & \text{if } x < 1 \end{cases}$$

Find formulas for the functions $g \circ f$ and $f \circ g$. Use this example to show that $g \circ f = f \circ g$ is not in general true.

(3) Define $f : \mathcal{R} \to \mathcal{R}$ and $g : \mathcal{R} \to \mathcal{R}$ by

$$f(x) = x^2$$
$$g(x) = \begin{cases} x + 1 & \text{if } x > 0 \\ -10 & \text{if } x \leq 0 \end{cases}$$

Find $g \circ f$ and $f \circ g$. In each case say whether the function is one-one, onto, both, or neither, and identify the ranges of each of the functions f, g, $f \circ g$, $g \circ f$.

(4) Define $f : \mathcal{R} \to \mathcal{R}$ by $f(x) = \sin x$ and $g : \mathcal{R} \to \mathcal{R}$ by $g(x) = 2x^2 + 1$. find $f \circ g$ and $g \circ f$, and identify the range in each case.

(5) Let $A = \{\{1, 2\}, \{1, 3\}, \{2\}, \{2, 5\}, \{3, 4, 5\}\}$. Define $f : A \to \mathcal{N}$ by

$$f(a) = \text{ the sum of the elements of } a \quad (\forall a \in A)$$

Find $f[A]$.
Now define $g : \mathcal{N} \to \mathcal{N}$ by

$$g(x) = \begin{cases} \text{the largest prime } \le n & \text{if } n > 1 \\ 1 & \text{if } n = 1 \end{cases}$$

Find $g \circ f$ and identify its range. Is g one-one? Is $g \circ f$ one-one?

(6) Let $f : A \to B$. Prove that

$$f \circ I_A = I_B \circ f = f$$

where I_A, I_B are the identity functions on A, B, respectively.

(7) Let $f : A \to B$ be invertible. Show that

$$f^{-1} \circ f = I_A \text{ and } f \circ f^{-1} = I_B$$

(8) Define $f : \mathcal{R} \to \mathcal{R}$ by

$$f(x) = \begin{cases} x^2 + 1 & \text{if } x \ge 0 \\ x + 1 & \text{if } x < 0 \end{cases}$$

Show that f is a bijection and find f^{-1}.

(9) Define $f : \mathcal{R}^2 \to \mathcal{R}^2$ by $f(x, y) = (x + 2y, x - y)$. Show that f is a bijection and find f^{-1}.

(10) Suppose both $f : A \to B$ and $g : B \to C$ are invertible. Show that $g \circ f$ is invertible and that $(g \circ f)^{-1} = f^{-1} \circ g^{-1}$.

(11) Let A be the set of all countries, B the set of all capital cities of the world. Define $f : A \to B$ by

$$f(a) = \text{ the capital of the country } a \quad (\forall a \in A)$$

Describe the inverse function f^{-1}.

(12) Let $f : A \to B$ be invertible. Prove that

$$f^{-1}(b) = \text{ the unique pre-image of } b \text{ under } f \quad (\forall b \in B)$$

(13) Let $f : A \to B$, $g : B \to C$. Prove that:

 (a) if f, g are one-one, then $g \circ f$ is one-one;

 (b) if f, g are onto, then $g \circ f$ is onto;

 (c) if $g \circ f : A \to C$ is onto, then g is onto.

(14) Let $f : A \to B$ be a bijection. Prove that f^{-1} is a bijection.

4.6 Denumerability

A particularly fascinating application of the abstract function concept is the way it provides us with a means of dealing with infinite sets, sets such as the set of all natural numbers or the set of all negative reals. Alhough our present level of abstraction might make our account seem a bit daunting at first, the basic idea is really as old as the notion of counting itself; indeed, *it is* that very notion of counting finite collections that we adopt in order to cope with infinite sets.

The fundamental notion that lies behind counting is that of *pairing off*. Early societies, and some present-day primitive peoples, would keep a record of "how many" objects of a particular kind were owned by some form of physical tokens, such as pebbles or small sticks. For each goat owned, say, the owner would have a pebble in the pile, and if asked how many goats were owned, the reply would be to point to the pile of pebbles and say "that many". For such a simple purpose, there is no need for a universally adopted system of abstract *numbers*. It is enough that the pebbles in the pile can be paired off, one-to-one, with the goats owned. Put crudely, this form of counting does not require *numbers*. But it does assume the notion of a one-to-one correspondence, what we would nowadays call a bijection. (Not that primitive peoples would have thought of it in this way!)

Nowadays we don't use piles of pebbles to keep track of collections any more, or at least we don't do so very often. We have the counting numbers — the natural numbers — which we can think of as abstract analogues of those pebbles. If we mechanically count the elements of some (finite) collection A of objects, "one, two, three", etc., what we are doing is explicitly exhibiting a one-to-one correspondence (a bijection) between a set of natual numbers and the elements of A. That is to say, we may picture the counting process as follows:

$$
\begin{array}{ccccccc}
a_1 & a_2 & a_3 & \cdots & \cdots & a_{n-1} & a_n \\
\uparrow & \uparrow & \uparrow & & & \uparrow & \uparrow \\
1 & 2 & 3 & \cdots & \cdots & n-1 & n
\end{array}
$$

Thus, a_1 is the *first* element of A (according to our count), a_2 is the *second*, a_3 the *third*, and so on. As the above diagram shows, this counting process determines a bijection between the set $\{1, 2, 3, \ldots, n-1, n\}$ and A. Since the counting process stops when we get to the number n (i.e., when we reach the nth element of A), we say that the set A "has n elements", or that "the number of elements of A is n".

So far this must all seem to be making a lot of fuss about something very familiar. So, while I am already running the risk of putting you to sleep, let me make two more very obvious observations. First, the number we reach when we count the elements in a given (finite) set does not depend on the order in which we count the elements. Second, given enough time we could carry out such a counting process for any finite set whatsoever.

In fact, our second observation really amounts to a definition of what it means for a set to be finite; namely a set A is finite iff the process of counting its members comes to an end. More formally, a set A is *finite* iff, for some natural number n, there is a bijection from the set $\{1, 2, 3, \ldots, n - 1, n\}$ to A.

Now let's turn our attention from finite sets to infinite ones. As Cantor (whom we met in Section 3.8) observed at the turn of the century, the same, age-old idea of pairing off that we have just discussed can also be used to count the members of an infinite set. In fact what Cantor did was take this simple idea and use it to develop an entire system of infinite numbers, having its own form of arithmetic. We shall not delve very far into that theory here; rather we'll just look at one particular aspect that turns out to be very important in a number of branches of mathematics: the related notions of denumerability and uncountability.

An infinite set A is said to be *denumerable* if there is a bijection

$$f : \mathcal{N} \to A$$

That is to say, A is denumerable iff it can be put into one-to-one correspondence with the set of all natural numbers. Let's look at a few simple examples.

- The set \mathcal{N} is itself denumerable; the identity function on \mathcal{N} is a suitable bijection.

- Let E be the set of even natural numbers. Then E is denumerable. The function $f : \mathcal{N} \to E$ defined by

$$f(n) = 2n$$

is a bijection. Similarly, the function $g(n) = 2n - 1$ is a bijection from \mathcal{N} to the set of all odd natural numbers, so that set too is denumerable.

- The set $R = \{1/n \mid n \in \mathcal{N}\}$ is denumerable. Consider the function $f : \mathcal{N} \to R$ defined by

$$f(n) = 1/n$$

- The set, \mathcal{Z}^- of all negative integers is denumerable. Consider the function defined on \mathcal{N} by $f(n) = -n$.

- The set \mathcal{Z} of all integers is denumerable. To see this, define a function $f : \mathcal{N} \to \mathcal{Z}$ by:

$$f(n) = \begin{cases} n/2, & \text{if } n \text{ is even} \\ -(n-1)/2, & \text{if } n \text{ is odd} \end{cases}$$

- The set P of all prime numbers is denumerable. To see this, we define a function $f : \mathcal{N} \to P$ by the following definition: let $f(1) = 2$, and for any $n \geq 1$, let $f(n + 1)$ be the least prime number bigger than $f(n)$. First of all, you should convince yourself that this does in fact define a function. Then you should check that f is a bijection.

The following lemma generalizes this last example.

LEMMA 4.6.1
If A is any infinite set of natural numbers, then A is denumerable.

PROOF Define $f : \mathcal{N} \to A$ by letting $f(1)$ be the smallest number in A and, for any $n \geq 1$,

$$f(n + 1) = \text{the smallest number in } A \text{ that is greater than } f(n).$$

As in the last example above, we have to prove that this defines a function, and that f is a bijection from \mathcal{N} to A. These are left as exercises for the reader. ⬜

LEMMA 4.6.2
If T is denumerable and there is a bijection $f : T \to A$ from T to some other set A, then A is denumerable.

PROOF Let $h : \mathcal{N} \to T$ be a bijection. Then the composition function $f \circ h$ is a bijection from \mathcal{N} to B. ⬜

LEMMA 4.6.3
An infinite subset of a denumerable set is denumerable.

PROOF Let A be denumerable, B an infinite subset of A, $f : \mathcal{N} \to A$ a bijection that testifies to the denumerability of A. Let

$$T = \{n \in \mathcal{N} \mid f(n) \in B\}$$

Being an infinite subset of \mathcal{N}, T is denumerable, by Lemma 4.6.1. But clearly, the function f provides a one-to-one correspondence between T and B. So by Lemma 4.6.2, B is denumerable. ⬜

LEMMA 4.6.4
The set $\mathcal{N} \times \mathcal{N}$ is denumerable.

PROOF The function $f : \mathcal{N} \times \mathcal{N} \to \mathcal{N}$ defined by

$$f(m, n) = 2^{m-1}(2n - 1)$$

is a bijection. (Exercise: prove this fact.) Hence its inverse function f^{-1} exists and is also a bijection. This inverse function testifies to the denumerability of $\mathcal{N} \times \mathcal{N}$. □

For our final result we prove that the set \mathcal{Q}^+ of all positive rational numbers is denumerable. This is a somewhat surprising result, since the mental picture we have of the positive real-number line has the natural numbers spaced out at regular, discrete intervals, with infinitely many rational numbers between any pair of natural numbers — indeed, there are infinitely many rational numbers between any two rational numbers. This picture would seem quite at odds with there being a way to pair off the positive rational numbers with the natural numbers. But the world of infinite sets is full of counter-intuitive surprises of this kind.

LEMMA 4.6.5
The set \mathcal{Q}^+ is denumerable.

PROOF Every positive rational number q can be written uniquely in the form m_q/n_q where m_q, n_q are positive integers having no common factor. Thus the function

$$f : \mathcal{Q}^+ \to \mathcal{N} \times \mathcal{N}$$

defined by $f(q) = (m_q, n_q)$ is one-one. Let T be the range of f. T is an infinite subset of $\mathcal{N} \times \mathcal{N}$, so by Lemmas 4.6.4 and 4.6.3, T is denumerable. But f determines a bijection from \mathcal{Q}^+ onto T, and the inverse of this bijection is a bijection from T onto \mathcal{Q}^+. Hence by Lemma 4.6.2, \mathcal{Q}^+ is denumerable. □

Exercises 4.6

(1) Show that the set of all positive multiples of 3 is denumerable.

(2) Show that the set $\{n \in \mathcal{N} \mid n \geq 10\}$ is denumerable.

(3) Show that the set $\mathcal{N} - \{4, 5\}$ is denumerable.

(4) Prove that if A is denumerable and x is an object not in A, then $A \cup \{x\}$ is denumerable.

(5) Prove that if A and B are disjoint denumerable sets, then $A \cup B$ is denumerable.

(6) Prove that if A and B are denumerable sets, not necessarily disjoint, then $A \cup B$ is denumerable.

(7) Prove that the set of *all* rationals is denumerable.

(8) Prove that if there is a one-one function from a set A into \mathcal{N}, then A is either finite or denumerable. (The term *countable* is often used to describe a set that

is either finite or denumerable.) Give an example to show that the assumption that the function be one-one cannot be omitted here.

(9) Prove that if there is a function from \mathcal{N} *onto* a set A, then A is countable (i.e., either finite or denumerable).

(10) Fill in the details in the proof of Lemma 4.6.4. In particular, prove that the function f defined there is a bijection.

4.7 Uncountability

Most students will find this section considerably more difficult than the other material in this book, and you might wish to skip it on first reading.

In the previous section we observed that a great many infinite sets are denumerable, and the reader might have begun to wonder if it were not in fact the case that every infinite set is denumerable. Following Cantor, we show that this is not true, and indeed, the set \mathcal{R} of real numbers is not denumerable.

Since the word *countable* is used to describe a set that is either finite or denumerable, an *uncountable* set is one that is infinite and non-denumerable. Thus Cantor's result shows that the set of reals is uncountable.

Before we start the proof, I should stress that although the argument is fairly short, it has considerable logical sophistication, and it may take you some considerable time, and several rereadings of the proof, in order to understand what is going on. After I have finished giving the formal proof, I shall try to provide some explanation to help clarify the reasoning.

THEOREM 4.7.1 (Cantor)
The set \mathcal{R} is non-denumerable.

PROOF Suppose, on the contrary, that \mathcal{R} were denumerable. Then by Lemma 4.6.3, the unit interval $(0, 1]$ would be denumerable. Let $f : \mathcal{N} \to (0, 1]$ be a bijection that testifies this fact.

Now, each number in $(0, 1]$ can be represented in a unique way as a non-terminating decimal, i.e., a decimal that has infinitely many non-zero entries. [In the case of rational numbers having a finite decimal expansion, there is an alternative decimal expansion that is infinite. You obtain this by reducing the final decimal by 1 and then writing an infinite string of 9's. For example, $0.45 = 0.44999\ldots$.]

For each natural number n, denote the non-terminating decimal representation of the number $f(n)$ by:

$$f(n) = 0.d_1^n \, d_2^n \, d_3^n \cdots d_i^n \cdots$$

Thus d_i^n denotes the digit in the ith decimal place in the decimal representation of $f(n)$ as a non-terminating decimal.

We now specify a new number r. We specify r by giving its decimal expansion; that is to say, we describe a rule that determines the decimal expansion of r. The number r has the decimal representation

$$r = 0.r_1 r_2 r_3 \cdots r_i \cdots$$

where for every index $i = 1, 2, 3, \ldots$ we define

$$r_i = \begin{cases} 1, & \text{if } d_i^i \neq 1 \\ 2, & \text{if } d_i^i = 1 \end{cases}$$

This clearly specifies a real number in the unit interval $(0, 1]$.

Now, since f is onto $(0, 1]$, there must be some natural number k such that $r = f(k)$. In particular, r and $f(k)$ must have the same decimal representation; that is to say, it must be the case that for every index i,

$$r_i = d_i^k$$

In particular, this must hold for the index $i = k$, that is to say,

$$r_k = d_k^k$$

But by definition of r, if $d_k^k = 1$ then $r_k = 2$ and if $d_k^k \neq 1$ then $r_k = 1$. Hence, in any event

$$r_k \neq d_k^k$$

This contradiction proves the theorem. ⬚

The argument presented above is generally referred to as Cantor's *diagonal argument*. As I mentioned in Section 3.8, it was his second proof of the result. To see how the name originates, suppose we present the proof as follows. Assuming that the real interval $(0, 1]$ is denumerable, we can list all the members of this interval as $f(1), f(2), f(3), \ldots$ Now let's write out the decimal representations of these numbers in a regular way, namely:

$$
\begin{array}{lllllll}
f(1) = 0. & d_1^1 & d_2^1 & d_3^1 & d_4^1 & d_5^1 & \cdots \quad \cdots \\
f(2) = 0. & d_1^2 & d_2^2 & d_3^2 & d_4^2 & d_5^2 & \cdots \quad \cdots \\
f(3) = 0. & d_1^3 & d_2^3 & d_3^3 & d_4^3 & d_5^3 & \cdots \quad \cdots \\
f(4) = 0. & d_1^4 & d_2^4 & d_3^4 & d_4^4 & d_5^4 & \cdots \quad \cdots \\
f(5) = 0. & d_1^5 & d_2^5 & d_3^5 & d_4^5 & d_5^5 & \cdots \quad \cdots \\
\vdots & \vdots & \vdots & \vdots & \vdots & \vdots & \vdots \quad \vdots
\end{array}
$$

Remember that this list is, following our initial assumption, a listing of *all* the reals in the interval (0, 1]. Thus, if we can define a real number r that lies between 0 and 1 but is not equal to any member of this list, we shall have a contradiction, and this will demonstrate that the original assumption is false, that is to say, the set \mathcal{R} must in fact be non-denumerable. The way we obtain such a number r is by specifying its decimal expansion. And we do this by working our way down the main *diagonal* of the above array, as follows.

$$
\begin{array}{llllllll}
f(1) = 0. & \boxed{d_1^1} & d_2^1 & d_3^1 & d_4^1 & d_5^1 & \cdots & \cdots \\
f(2) = 0. & d_1^2 & \boxed{d_2^2} & d_3^2 & d_4^2 & d_5^2 & \cdots & \cdots \\
f(3) = 0. & d_1^3 & d_2^3 & \boxed{d_3^3} & d_4^3 & d_5^3 & \cdots & \cdots \\
f(4) = 0. & d_1^4 & d_2^4 & d_3^4 & \boxed{d_4^4} & d_5^4 & \cdots & \cdots \\
f(5) = 0. & d_1^5 & d_2^5 & d_3^5 & d_4^5 & \boxed{d_5^5} & \cdots & \cdots \\
& \vdots & \vdots & \vdots & \vdots & \vdots & \vdots & \vdots
\end{array}
$$

To determine the first decimal digit of r, we look at the first digit in the diagonal, d_1^1. If this digit is a 1, we let the first digit of r be 2; if the first diagonal digit is not 1, we let the first digit of r be 1. Thus, however we specify the remainder of the decimal expansion of r, we have ensured that r cannot be equal to $f(1)$, since r and $f(1)$ differ in the first decimal place.

To specify the second decimal digit of r, we look at the second entry in the diagonal, the digit d_2^2. The same way as before, we define the second digit of r to be different from the second digit of $f(2)$, so that now, however we define the remainder of the expansion of r, r will not be equal to $f(2)$.

And so on. At stage i in this definition, we make sure that the ith decimal digit of r is different from d_i^i, the ith decimal digit of $f(i)$.

Thus for every index i, r cannot be equal to $f(i)$, since r and $f(i)$ have different entries in the ith decimal place.

Cantor's theorem tells us that there are at least two different sizes of infinite sets, the denumerable and the non-denumerable. In fact, by generalizing the Cantor diagonalization argument, it is possible to show, as Cantor himself did, that there is in fact an entire infinite progression of increasing infinities. But for the bulk of mathematics, it is the countable versus uncountable distinction that is important. So in this book we shall simply take a brief look at the world of uncountable sets insofar as they compare in size to the set of reals.

Two sets A, B are said to be *equipollent* iff there is a bijection $f : A \to B$. Clearly, in the case of finite sets, A and B will be equipollent iff they have the same number of elements. (Exercise: prove this fact.) And by definition, an infinite set A will be denumerable iff it is equipollent to the set \mathcal{N}. That leaves the issue of infinite sets equipollent to \mathcal{R}. Any such set is very definitely "bigger" (in an infinite sense) than the sets \mathcal{N}, \mathcal{Z}, \mathcal{Q}. We give a couple of simple but important examples.

- The unit interval $(0, 1)$ is equipollent to the entire real line \mathcal{R}. To see this, consider the bijection $f : (0, 1) \rightarrow \mathcal{R}$ defined by

$$f(x) = \tan(\pi x - \pi/2)$$

(Exercise: sketch the graph of this function to see how it works here.)

- Any two non-empty real intervals (a, b), (c, d) are equipollent. Take the bijection $f : (a, b) \rightarrow (c, d)$ defined by

$$f(x) = \frac{d - c}{b - a} x + \frac{bc - ad}{b - a}$$

Exercises 4.7

(1) Let B be the set of all infinite binary sequences (i.e., all infinite sequences of 0's and 1's). Prove that B is uncountable.

(2) Prove that if A is uncountable and B is a countable subset of A, then $A - B$ is uncountable. (It is in fact equipollent to A, though you are not being asked to prove this additional fact.)

(3) Prove that if A and B are disjoint sets equipollent to \mathcal{R}, then $A \cup B$ is equipollent to \mathcal{R}. (The conclusion is in fact valid even without the assumption of disjointness.)

(4) Prove that the set of irrational numbers is uncountable. (It is in fact equipollent to \mathcal{R}, though you are not being asked to prove this.)

(5) Prove that if X is an infinite set, the power set of X, $\mathcal{P}(X)$, is not equipollent with X, and thus its size is a larger class of infinity. (Do not make any assumptions about X other than it is infinite.)

Chapter 5

Relations

5.1 Binary Relations

The notion of a *binary relation* (which we shall refer to simply as a *relation* for most of this chapter) is one that will be familiar to you. Given sets A and B, a *relation between* A and B links certain elements of A with certain elements of B.

For example, if A is the set of all people and B is the set of all objects, then *ownership* is a relation between A and B. In this case, a person p is *related* to an object a iff p *owns* a.

Again, if U is a universal set and A is a collection of subsets of U, the membership relation \in is a relation between U and A.

In mathematical discussions we generally refer to a relation between the sets A and B in terms of the Cartesian product, namely as a relation *on* $A \times B$. The reasons for this nomenclature will become clear in due course. In the meantime, just put it down to the usual eccentricities of mathematicians.

Most of the examples of relations that arise in mathematics (and many of the relations that arise in everyday life) are relations that connect elements of the same set; that is to say, they are relations on $A \times A$ for the appropriate set A. We refer to such relations simply as relations *on* A.

For example, $<, \leq, =$ are all relations on the set \mathcal{N} of natural numbers. In the case of the less-than relation $<$, given any pair m, n of natural numbers, either it is the case that $m < n$ or else it is not the case that $m < n$. In case $m < n$ we say that m, n are related by the less-than relation, or that they *satisfy* this relation; in this case the proposition $m < n$ expresses a certain *relation*ship between m and n, namely that m is *less than* n. Likewise for the relations \leq and $=$.

For another example, if A is any collection of sets, \subseteq is a relation on A. Given two sets X, Y in A, either $X \subseteq Y$ or else $X \nsubseteq Y$.

Again, the relation of *disjointness* on A: two sets X, Y are *disjoint* iff $X \cap Y = \emptyset$. There is no standard symbol for this relation as there is for the preceding examples. Usually one just writes $X \cap Y = \emptyset$ or $X \cap Y \neq \emptyset$. But this is still a *relation*ship between two sets; the role played by the empty set is purely to express the way the two sets are or are not related, namely by having no/some elements in common.

Finally, if P is the set of all people, there is the relationship of marriage. Certain pairs of people are married and other pairs are not married.

Alhough the concept of a relation is certainly one you have met many times, both in mathematics and in your everyday activities, what you might well find novel is to regard relations as mathematical *objects* of study in their own right, rather than as concepts that relate the objects under study.

In order to carry out a systematic study of relations, we introduce a little terminology and some notational conventions.

If R is a relation on $A \times B$, we call the set A the *domain* of R and B the *codomain* of R.

In the case where R is a relation for which the domain and codomain are the same set A, that set A is called the *underlying set* for R.

In mathematical studies, in order to specify a relation, the domain and codomain (or the underlying set where appropriate) must be specified.

This implies, for example, that there is not one less-than relation but several, one on \mathcal{N}, one on \mathcal{Z}, one on \mathcal{Q}, one on \mathcal{R}, one on the set of primes, one on the set of all irrational numbers, etc. Of course, there is a sense in which these are all the same relation: the relation of one number being *less than* another. But for the purposes of the kind of study we wish (and need) to make here, it is important that we distinguish between all of these different instances of the less-than relationship that arise by taking different underlying sets. The reason for being specific about underlying sets will become clear in the next section when we examine the distinctions that can be made between different types of relations.

In general discussions of relations, we often use the letters R, S to denote arbitrary relations. In keeping with the common notation adopted for relations such as $<, \leq, =$, we use what is known as *infix* notation, whereby the proposition that the elements a, b of the domain and codomain satisfy the relation R is written in the form

$$a\,R\,b$$

Alternative notations that are possible include Rab (*prefix* notation), abR (*postfix* notation), and $R(a, b)$.

Here are some further examples of relations:

- R on \mathcal{N} defined by: $m\,R\,n$ iff $m > 10n$;

- R on \mathcal{Z} defined by: $m\,R\,n$ iff $m + n$ is even;

- R on \mathcal{Z} defined by: $m\,R\,n$ iff $m - n$ is even;

- E on \mathcal{R} defined by: $x\,E\,y$ iff $y = e^x$;

- R on \mathcal{R} defined by $x\,R\,y$ iff $xy \geq 0$;

- S on the set, T, of all triangles in the plane defined by $t_1\,S\,t_2$ iff t_1 and t_2 are similar;

- \perp (perpendicularity) on the set, L, of all straight lines in the plane;

- R with domain \mathcal{R} and codomain the set, P, of all real polynomials in one unknown, defined by aRp iff a is a root of p;

- R on the set $P \times L$, where P is the set of all points in the plane, L is the set of all straight lines in the plane, and pRl means that the point p lies on the line l;

- P on the set, A, of all people, where xPy means x is a parent of y;

- S on the set of all Americans where aSb means a is a sibling of b;

- B on the set of all Americans where aBb means a is a brother of b;

- R on the set \mathcal{R} defined by xRy iff $(x < 0)$ or $(x \geq 0 \wedge x = y)$.

Exercises 5.1

(1) Of the examples of relations given in the chapter so far, which ones differentiate between the order of the objects related, and for which ones is the order of the elements unimportant? (This question is only relevant to relations where the domain and codomain are the same.)

(2) For each of the examples given so far, which ones relate objects to themselves and which do not? (Again, this only applies to relations where the domain equals the codomain.)

(3) Give three further examples of relations where the domain and codomain are different.

(4) Give six further examples of relations where the domain and codomain are the same, three where the underlying set is mathematical, three where the underlying set is a set of people.

5.2 Properties of Relations

From now on, unless specified otherwise, we restrict our attention to relations where the domain and codomain are equal.

Let R be a relation on a set A. We say R is *reflexive* iff

$$(\forall a \in A)[aRa]$$

For example, the relation \leq on \mathcal{R} is reflexive: for every real x, it is the case that $x \leq x$. On the other hand, the relation $<$ on \mathcal{R} is not reflexive. Indeed, for *no* real

x is it the case that $x < x$. This is far more than is required for reflexivity to fail. Since reflexivity is defined using a universal quantifier, it only requires one counter-example in order for reflexivity to fail.

For example, consider the relation R on \mathcal{R} defined by

$$x R y \text{ iff } xy > 0$$

For most reals x, $x^2 > 0$, so $x R x$. In fact there is only one real x for which this fails, and that is 0. But because $\neg(0R0)$, this relation is not reflexive. If we change the underlying set from \mathcal{R} to the set of all non-zero reals, then the resulting relation *is* reflexive.

This last example provides an indication of the importance of stipulating the underlying set of a relation.

We say a relation R on a set A is *symmetric* iff

$$(\forall a, b \in A)[a R b \rightarrow b R a]$$

Notice that this definition immediately implies a stronger form of itself where \Rightarrow is replaced by \Leftrightarrow. Thus what symmetry says is that the order of the elements is not important as far as the relation is concerned.

For example, the relation R on \mathcal{N} defined by

$$m R n \text{ iff } m - n \text{ is even}$$

is symmetric: for any two natural numbers m, n, if $m - n$ is even then $n - m$ is even, i.e., $m R n$ implies $n R m$.

On the other hand, the relation $<$ on \mathcal{Z} is not symmetric. Indeed, there are *no* pairs m, n of integers such that both $m < n$ and $n < m$, so certainly it is not the case that

$$(\forall m, n \in \mathcal{Z})[m < n \rightarrow n < m]$$

Notice once again that in order for symmetry to fail it is enough for there to be just one pair of elements of the underlying set for which the symmetry property fails.

For example, define a relation R on \mathcal{R} by:

$$x R y \text{ iff } (x = 0 \wedge y = 1) \vee (x \neq 0 \wedge x = y)$$

Then (and you should check this for yourself) we have $[x R y \rightarrow y R x]$ for all pairs x, y of reals except $x = 0, y = 1$. The existence of this one counter-example is enough to make R non-symmetric. If we change the underlying set to be the set of all non-zero reals, then the resulting relation is symmetric.

A relation R on a set A is said to be *transitive* iff

$$(\forall a, b, c \in A)[a R b \wedge b R c \rightarrow a R c]$$

For example, the relation $<$ on \mathcal{Z} is transitive: if $x < y$ and $y < z$, then $x < z$.

The relation \perp of perpendicularity on the set, L, of straight lines in the plane is not transitive. Indeed, if $p \perp q$ and $q \perp r$, then r and p must be parallel, so the transitivity condition fails in all possible cases.

Once again this kind of failure is extreme; a single counter-example is enough to cause failure. For example, define a relation R on \mathcal{Z} by

$$x R y \text{ iff } (x = 0 \wedge y = 1) \vee (x = 1 \vee y = 0) \vee (x \neq 0 \wedge x \neq 1 \wedge x = y)$$

Then we have $x R y \wedge y R z \to x R z$ for all values of x, y, z except those involving 0 or 1. It is left as an exercise for you, the reader, to figure out exactly what happens when one or more of x, y, z is equal to 0 or 1.

Exercises 5.2

For each of the following relations, say which of the properties of reflexivity, symmetry, and transitivity are satisfied and which fail.

(1) $<$ on \mathcal{Z}.

(2) \leq on \mathcal{Z}.

(3) $=$ on \mathcal{R}.

(4) \subseteq on the collection, $\mathcal{P}(\mathcal{N})$, of all subsets of \mathcal{N}.

(5) \subset on the collection, $\mathcal{P}(\mathcal{N})$, of all subsets of \mathcal{N}.

(6) Disjointness on $\mathcal{P}(\mathcal{N})$.

(7) R on \mathcal{N} defined by: $m R n$ iff $m > 10n$.

(8) R on \mathcal{Z} defined by: $m R n$ iff $m + n$ is even.

(9) R on \mathcal{Z} defined by: $m R n$ iff $m - n$ is even.

(10) E on \mathcal{R} defined by: $x E y$ iff $y = e^x$.

(11) R on \mathcal{R} defined by $x R y$ iff $xy \geq 0$.

(12) S on the set, T, of all triangles in the plane defined by $t_1 S t_2$ iff t_1 and t_2 are similar.

(13) \perp (perpendicularity) on the set, L, of all straight lines in the plane.

(14) R with domain \mathcal{R} and codomain the set, P, of all real polynomials in one unknown, defined by $a R p$ iff a is a root of p.

(15) R on the set $P \times L$, where P is the set of all points in the plane, L is the set of all straight lines in the plane, and $p R l$ means that the point p lies on the line l.

(16) P on the set, A, of all people, where $x P y$ means x is a parent of y.

(17) S on the set of all Americans where aSb means a is a sibling of b.

(18) B on the set of all Americans where aBb means a is a brother of b.

(19) R on the set \mathcal{R} defined by xRy iff $(x < 0)$ or $(x \geq 0 \wedge x = y)$.

5.3 Relations as Sets of Ordered Pairs

Although we have tried to regard relations as mathematical *objects* in their own right, they have been somewhat tenuous objects, plucked from our intuitions of properties. Most people find a degree of tension in the use of the word "thing" in the phrase "a relation is a thing that links objects together."

In mathematics, it is common to objectify abstractions of this sort by means of a formal mathematical definition of a *model*. The model that is usually taken for a relation is as a set of ordered pairs. If R is a relation between the two sets A and B, we model R by means of the set:

$$\hat{R} = \{(a, b) \in A \times B \mid aRb\}$$

Thus, for any elements $a \in A$, $b \in B$,

$$aRb \text{ iff } (a, b) \in \hat{R}$$

Thus, the question as to whether or not aRb for a given pair a, b can be replaced by the set-membership question as to whether or not $(a, b) \in \hat{R}$.

Now there are a great many questions one can ask about relations. Is it a relation used by people, and if so what kind of people use it? Is it a legal relation, and if so under what legal system? Is it a purely mathematical relation, and if so who invented it? And so on. But for most mathematical applications, none of these questions is important. In mathematics, the only question of significance as far as relations are concerned is: which pairs of elements satisfy the relation and which do not? But this is precisely the question that can be answered in straight set-theoretic terms by looking at the model set \hat{R}.

In other words, as far as mathematics is concerned, all that you need is the set \hat{R} of those pairs of objects that are linked by the relation R. In other words, you don't need (real) *relations* at all, it is enough to have the corresponding *sets*.

Accordingly, it is common in mathematics to "forget" that relations are *relations*, and to think of them instead as sets of ordered pairs. That is to say, we refer to the model sets \hat{R} as "relations" and forget about R altogether. This gives us the following alternative definition:

A *relation* between a set A and a set B is a subset of the Cartesian product $A \times B$.

In particular, a relation on a set A is a subset of $A \times A$.

Having made the switch from real relations to their set-theoretic counterparts, we now have no further need for the notation \hat{R}, of course. We can just speak about "the relation R," it now being understood that R will be a set of ordered pairs.

In terms of our new definition, the properties of reflexivity, symmetry, and transitivity for a relation R on a set A now read as follows:

- R is reflexive iff

$$(\forall a \in A)[(a, a) \in R]$$

- R is symmetric iff

$$(\forall a, b \in A)[(a, b) \in R \to (b, a) \in R]$$

- R is transitive iff

$$(\forall a, b, c \in A)[((a, b) \in R \wedge (b, c) \in R) \to ((a, c) \in R)]$$

At least, this is the new official way of writing these definitions. In practice, we continue to use our previous, less cumbersome infix notation, the only difference being that this is now just an abbreviation. That is to say, we agree to write aRb in place of $(a, b) \in R$.

This takes us back to our starting point, except that now we no longer have to accept relations as some kind of strange objects; they are just sets — specifically, subsets of Cartesian products.

In fact we now have far more freedom than before in constructing relations. According to our new definition, *any* subset of $A \times B$ will be a relation between A and B, regardless of whether or not there is some nice way to describe this relation.

For example, we may define a relation by simply listing the ordered pairs it contains, such as the relation R between the sets $\{1, 3, 5, 7, 9\}$ and $\{2, 4, 6, 8\}$ defined by:

$$R = \{(1, 2), (3, 2), (9, 2), (3, 6), (3, 8), (7, 2), (7, 4), (7, 6), (7, 8)\}$$

This issue is taken up again in the exercises for this section.

It should be noted that the above procedure of replacing a rather abstract notion by a formal mathematical model is a common one in mathematics. In fact we already met this device in Chapter 2, when we replaced the complex notion of implication by the formally defined (material) conditional \Rightarrow. Just as the conditional captures something, but not everything, about implication, so too our definition of a relation as a set of ordered pairs captures something, but not everything, about real relations.

Thus, our (new) definition of, say, the relation of marriage as being simply the set

$$\{(a, b) \mid a \text{ is married to } b\}$$

certainly does not capture everything about the relationship of marriage; but it does capture enough for our present, mathematical purposes.

Exercises 5.3

From now on, we always assume that a relation is a set of ordered pairs.

(1) Let $A = \{1, 2, 3\}$. Define (by explicitly writing down the set of ordered pairs) a relation on A that is:

 (a) reflexive, not symmetric, not transitive;

 (b) not reflexive, symmetric, not transitive;

 (c) not reflexive, not symmetric, transitive;

 (d) not reflexive, symmetric, transitive;

 (e) reflexive, not symmetric, transitive;

 (f) reflexive, symmetric, not transitive;

 (g) not reflexive, not symmetric, not transitive;

 (h) reflexive, symmetric, transitive.

(2) Which of the properties reflexive, symmetric, transitive do the following relations satisfy?

 (a) R on $\mathcal{R} \times \mathcal{R}$ defined by

$$(x, y)R(z, w) \text{ iff } x + z \le y + w$$

 (b) R on $\mathcal{R} \times \mathcal{R}$ defined by

$$(x, y)R(z, w) \text{ iff } x + y \le z + w$$

(3) Let R be a relation on A such that $(\forall a \in A)(\exists b \in A)[aRb]$. Show that if R is symmetric and transitive, then R is reflexive.

(4) Is it true that if R and S are relations on $A \times B$, then $R \cup S$ and $R \cap S$ are relations on $A \times B$? Justify your answer.

(5) Which of the following, if any, are true? (Justify your answer in each case.)

 (a) If R, S are reflexive relations on A, then $R \cup S$ is a reflexive relation on A.

 (b) If R, S are reflexive relations on A, then $R \cap S$ is a reflexive relation on A.

 (c) If R, S are symmetric relations on A, then $R \cup S$ is a symmetric relation on A.

 (d) If R, S are symmetric relations on A, then $R \cap S$ is a symmetric relation on A.

 (e) If R, S are transitive relations on A, then $R \cup S$ is a transitive relation on A.

(f) If R, S are transitive relations on A, then $R \cap S$ is a transitive relation on A.

(6) A relation R on a set A is said to be *asymmetric* iff

$$(\forall a, b \in A)[aRb \rightarrow \neg(bRa)]$$

and is said to be *antisymmetric* if

$$(\forall a, b \in A)[(aRb \wedge bRa) \rightarrow (a = b)]$$

Prove that if R is asymmetric, then it is antisymmetric. What can you say about the converse?

(7) A relation on A that is reflexive, antisymmetric, and transitive is called a *partial order* on A. Prove that each of the ordering relations $<$ and \leq are partial orders on \mathcal{R} and that \subset and \subseteq are partial orderings on $\mathcal{P}(X)$ for any set X.

5.4 Relations as Graphs

There are two notions of graph in mathematics, so it is best to be sure at the outset which notion is being referred to by the word here.

By a *graph* we mean a set V, whose elements will be referred to as the *vertices* (of the graph), together with a collection of *edges* that connect certain pairs of vertices. It is possible to have an edge that runs from a vertex to itself. The set of vertices may be finite or infinite.

If each edge of the graph has a designated direction, the graph is said to be a *directed graph* (or sometimes more simply a *digraph*). It is possible for an edge in a digraph to be assigned both directions.

For example, Figure 5.1 presents a (finite) directed graph whose vertex-set V is the set $\{a, b, c, d, e, f, g, h\}$. In this directed graph, a is connected to b, b to c, a to c, c to a, c to itself, and so on.

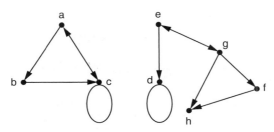

Figure 5.1: A directed graph.

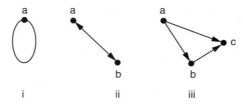

Figure 5.2: Reflexivity, symmetry, and transitivity.

Directed graphs provide a convenient way to illustrate relations, at least relations defined on a small, finite set of elements. This is somewhat analogous to the use of Venn diagrams to illustrate set-theoretic configurations. If R is a relation on a set A, we can represent R as a digraph that has A as vertex set and has an edge $a \longrightarrow b$ for each pair a, b from A such that aRb.

Figure 5.1 represents the relation on the set

$$A = \{a, b, c, d, e, f, g, h\}$$

whose ordered pair representation is:

$$R = \{(a, b), (b, c), (a, c), (c, a), (c, c), (d, d), (e, d),$$
$$(e, g), (g, e), (g, f), (f, h), (g, h)\}$$

Figure 5.2 indicates the graph-theoretic properties that correspond to the reflexivity, symmetry, and transitivity of relations. A reflexive relation is one whose digraph has the property that every vertex has a loop attached to it, as in Figure 5.2 (i). A symmetric relation is one whose digraph is such that every edge is directed both ways, as in Figure 5.2 (ii). A transitive relation is one in which every time there is a pair of directed edges $a \longrightarrow b$ and $b \longrightarrow c$, then there is an edge $a \longrightarrow c$ that completes the triangle, as in Figure 5.2 (iii).

We refer to the above properties of digraphs as *reflexivity, symmetry*, and *transitivity*, respectively.

5.5 Equivalence Relations

A relation R on a set A is said to be an *equivalence relation* if it is reflexive, symmetric, and transitive.

For example, each of the following is an equivalence relation:

- $=$ (equality) on any set A;

- R defined on \mathcal{Z} by mRn iff $m - n$ is even;

- R defined on \mathcal{N} by $(m, n)R(p, q)$ iff $mq = np$.

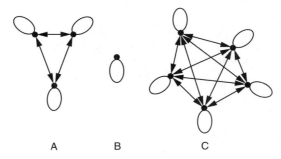

A B C

Figure 5.3: Graph of an equivalence relation.

The proof that each of these is an equivalence relation is left to the reader. In each case there are three conditions to verify, of course.

In the case of an equivalence relation, the associated digraph exhibits a very special form, illustrated by Figure 5.3. Namely, the graph divides up into distinct subgraphs, each of which is a reflexive, symmetric, transitive graph in which every pair of vertices is joined by an edge, and no two of which subgraphs are joined by any edge. In the example illustrated, these subgraphs are labeled A, B, C.

If R is an equivalence relation on a set A, then for any $a \in A$ we define the *equivalence class* of a (*modulo R*) to be the set:

$$[a] = \{x \in A \mid aRx\}$$

For example, let R be the equivalence relation defined on the set \mathcal{N} of natural numbers by mRn iff $m - n$ is even. Then, for instance,

$$[1] = \{n \mid mRn\} = \{n \mid 1-n \text{ is even}\} = \{n \mid n \text{ is odd}\}$$

and

$$[2] = \{n \mid mRn\} = \{n \mid 2-n \text{ is even}\} = \{n \mid n \text{ is even}\}$$

Continuing in this manner, we see that:

$$[1] = [3] = [5] = [7] = \ldots = \text{ the odd numbers}$$
$$[2] = [4] = [6] = [8] = \ldots = \text{ the even numbers}$$

Thus in this case there are just two equivalance classes, the set of even numbers and the set of odd numbers. The equivalence class of any odd number is the set of all odd numbers, and the equivalence class of any even number is the set of all even numbers. In particular, this particular equivalence relation splits the underlying set \mathcal{N} into two disjoint parts.

The following theorem shows that this is just a special case of a quite general phenomenon.

THEOREM 5.5.1

Let R be an equivalence relation on a set A.

 (i) *For any* $a \in A$, $a \in [a]$.

 (ii) *For any* $a, b \in A$, $a R b$ *iff* $[a] = [b]$.

 (iii) *For any* $a, b \in A$, *if* $\neg(a R b)$ *then* $[a] \cap [b] = \emptyset$.

 (iv) *Any two equivalence classes are either equal or disjoint.*

PROOF

 (i) Since R is reflexive, $a R a$, so

$$a \in \{x \in A \mid a R x\} = [a]$$

 (ii) Assume $[a] = [b]$. By (i), $b \in [b]$. Thus $b \in [a]$. That is,

$$b \in \{x \in A \mid a R x\}$$

Thus $a R b$.
 Conversely, assume $a R b$. We prove that $[a] = [b]$. Let $x \in [a]$ be arbitrary. Then by definition, $a R x$. Now by symmetry, our assumption gives $b R a$. Since $b R a$ and $a R x$, transitivity gives $b R x$. Thus $x \in [b]$. Now suppose $x \in [b]$. Then by definition, $b R x$. Thus we have $a R b$ and $b R x$, so by transitivity, $a R x$, giving $x \in [a]$. The proof of (ii) is complete.

 (iii) We prove the (logically equivalent) contrapositive, namely that if $[a] \cap [b] \neq \emptyset$ then $a R b$. Let $x \in [a] \cap [b]$. Since $x \in [a]$, $a R x$. Since $x \in [b]$, $b R x$, so by symmetry, $x R b$. Then $a R x$ and $x R b$, so by transitivity, $a R b$.

 (iv) For any pair $a, b \in A$, either $a R b$ or $\neg(a R b)$. If $a R b$ then by (ii), $[a] = [b]$. If $\neg(a R b)$, then by (iii), $[a] \cap [b] = \emptyset$. ☐

 By a *partition* of a set A we mean a collection \mathcal{F} of pairwise disjoint subsets of A (i.e., $X \cap Y = \emptyset$ for every distinct pair of sets X, Y in \mathcal{F}) whose union is the whole of A, i.e.,

$$A = \bigcup \mathcal{F}$$

 For example, if P is the set of all prime numbers, Q is the set of all composite numbers, and $R = \{1\}$, then the collection $\{P, Q, R\}$ is a partition of \mathcal{N}.
 Theorem 5.5.1 tells us that for any equivalence relation on a set A, the collection of all equivalence classes constitutes a partition of A. By part (i) of the theorem, A is equal to the union of all the equivalence classes. And by part (iv), the equivalence classes are pairwise disjoint.

This result has a converse, namely:

THEOREM 5.5.2
Let \mathcal{F} be a partition of a set A. Define a relation R on A by

$$a\,Rb \text{ iff } (\exists X \in \mathcal{F})[a, b \in X]$$

Then R is an equivalence relation on A. Moreover, the equivalence classes for R are precisely the sets in the partition \mathcal{F}.

PROOF Left as an exercise. ▯

The above observations tell us that equivalence relations are just a particular way of looking at an extremely familiar phenomenon: that of classification or categorization.

In many walks of life, as well as in mathematics, we frequently classify objects according to some criterion of equivalence. For instance, given a ranking of football teams or chess players, we often split them into *leagues* or *divisions* according to ability and performance. The equivalence relation involved here is one of "equivalent (or comparable) ability" and the equivalence classes are the different leagues or divisions.

Again, a number of U.S. publishers produce annual listings of American colleges ranked according to their degree of entrance selectivity. The equivalence relation here is one of "having roughly the same overall entrance requirements"; the equivalence classes are the groupings of colleges into categories (generally) known as "most competitive", "highly competitive", "very competitive", "competitive", "less competitive", and "non-competitive". A potential student who is told that two particular colleges are equally difficult to get into ($a\,Rb$) will know that they are in the same category ($[a] = [b]$); conversely, knowing that two colleges are both, say, ranked "highly competitive", (a, b in the same partition class) a potential student will know that they are equally hard to get into ($a\,Rb$).

The exercises below give some more mathematical examples of equivalence classes, and should provide some indication of the importance of this concept in modern mathematics.

Exercises 5.5

(1) Define R on \mathcal{N} by $m\,Rn$ iff 3 divides $m - n$.

 (a) Prove that R is an equivalence relation on \mathcal{N}.

 (b) What are [1], [2], [3]?

 (c) What are [4], [5], [6]?

 (d) How many equivalence classes are there altogether? Prove your answer.

(2) Define R on \mathcal{N} by $m\,Rn$ iff 3 divides $m+n$. Prove that R is not an equivalence relation on \mathcal{N}.

(3) Define R on \mathcal{R} by: $x\,Ry$ iff $x^2 = y^2$.
 (a) Prove that R is an equivalence relation on \mathcal{R}.
 (b) What are $[0]$, $[2]$, $[49]$?
 (c) Prove that there are infinitely many equivalence classes.

(4) Fix an integer $p \geq 1$, and define R on \mathcal{Z} by: $m\,Rn$ iff p divides $m-n$.
 (a) Prove that R is an equivalence relation on \mathcal{Z}.
 (b) What are $[0]$, $[1]$, $[-1]$, $[p]$, $[-p]$, $[p+1]$?
 (c) Prove that $[0], [1], \ldots, [p-1]$ are all distinct equivalence classes.
 (d) Prove that $[0], [1], \ldots, [p-1]$ are *all* the equivalence classes.

(5) Define R on $\mathcal{N} \times \mathcal{N}$ by:

$$(m, n)R(p, q) \text{ iff } mq = np$$

 (a) Prove that R is an equivalence relation on $\mathcal{N} \times \mathcal{N}$.
 (b) Give three members of each of the equivalence classes

$$[(1, 1)], [(1, 2)], [(2, 5)]$$

 (c) Give a complete description of the set $[(p, q)]$ for any pair p, q from \mathcal{N}.

(6) Let \mathcal{F} be the collection of all finite sets. Define \approx on \mathcal{F} by:

$$X \approx Y \text{ iff there is a bijection } f : X \to Y$$

 (a) Prove that \approx is an equivalence relation on \mathcal{F}.
 (b) What are the equivalence classes of the sets \emptyset, $\{\emptyset\}$, $\{1, 2, 3\}$?
 (c) What is the equivalence class of the set $\{a_1, \ldots, a_n\}$ for any distinct objects a_1, \ldots, a_n?
 (d) Show that there is a bijection from \mathcal{N} onto the collection, \mathcal{C}, of all equivalence classes.
 (e) Can you suggest a possible *formal definition* of the integers $0, 1, 2, \ldots$?

The relation \approx is known as *equipollence*. For a generalization to infinite sets, see Sections 4.6 and 4.7.

(7) Define R on $\mathcal{R} \times \mathcal{R}$ by: $(x, y)R(u, v)$ iff $3x - y = 3u - v$.
 (a) Prove that R is an equivalence relation on $\mathcal{R} \times \mathcal{R}$.
 (b) What are $[(4, 5)]$, $[(0, 0)]$?
 (c) Describe $[(a, b)]$ for any fixed values of a, b, both in set-theoretic terms and geometrically.

(8) Define R on $\mathcal{R} \times \mathcal{R}$ by $(x, y) R(u, v)$ iff $x^2 + y^2 = u^2 + v^2$.

 (a) Prove that R is an equivalence relation on $\mathcal{R} \times \mathcal{R}$.

 (b) What are $[(0, 0)]$, $[(1, 1)]$, $[(3, 4)]$?

 (c) Describe $[(a, b)]$ for any fixed values of a, b, both in set-theoretic terms and geometrically.

(9) Prove Theorem 5.5.2.

(10) Describe the equivalence relation on \mathcal{Z} determined by the partition $\{A, B\}$, where $A = \{x \mid x \leq 0\}$, $B = \{x \mid x > 0\}$.

(11) Describe, as sets of ordered pairs, the equivalence relations on the set $A = \{1, 2, 3, 4, 5\}$ associated with the following partitions:

 (a) $\{\{1, 2\}, \{3, 4, 5\}\}$

 (b) $\{\{1\}, \{2\}, \{3, 4, 5\}\}$

 (c) $\{\{1, 2, 3, 4\}, \{5\}\}$

5.6 Functions as Relations

In Chapter 4, we adopted a definition of a function that concentrated on the assignment of value to argument, rather than the particular rule used to specify that process. A direct consequence of this approach is that we may regard any function as a simple set of ordered pairs, the pairs that link arguments with their corresponding values.

More precisely, given a function $f : A \to B$, we can identify f with the set

$$\hat{f} = \{(a, b) \in A \times B \mid f(a) = b\}$$

The characteristic property of any function is that for any argument there is exactly one value associated with that argument. In terms of the set \hat{f}, this says

$$(\forall a \in A)(\exists! b \in B)[(a, b) \in \hat{f}]$$

We may thus *define* a function f from a domain A to a codomain B to be a subset of $A \times B$ such that

$$(*) \qquad\qquad (\forall a \in A)(\exists! b \in B)[(a, b) \in f]$$

Many authors do in fact use this as their basic definition of function. Although this is a convenient device for the mathematician, it is, however, somewhat misleading. Functions are not sets of ordered pairs, any more than relations are sets of ordered pairs. Functions are functions and relations are relations. What certainly is the case

is that functions may be *modeled* as sets of ordered pairs, just as relations may be so modeled. The difference between the set-theoretic model of a function and that of a relation is that the function must satisfy the additional requirement (∗).

In the sense of these common mathematical modeling devices then, we may speak of a *function* from a set A to a set B as "being" a *relation* f on $A \times B$ that satisfies the condition (∗).

For example, the function $f : \mathcal{R} \to \mathcal{R}$ defined by $f(x) = x^2$ can be defined to be the relation

$$\{(x, y) \in \mathcal{R} \times \mathcal{R} \mid y = x^2\}$$

Exercises 5.6

(1) Which of the following relations are functions and which are not? For those that are functions, indicate the domain and a possible codomain.

 (a) $R = \{(1, 2), (1, 3), (2, 4), (3, 4), (4, 5), (5, 5)\}$

 (b) $R = \{(1, 2), (2, 1)\}$

 (c) $R = \{(x, y) \mid x \in \mathcal{R}, y \in \mathcal{R}, x = \sin y\}$

 (d) $R = \{(m, n) \in \mathcal{N} \times \mathcal{N} \mid m^2 = n\}$

 (e) $R = \{(m, n) \in \mathcal{Z} \times \mathcal{Z} \mid m^2 = n\}$

 (f) $R = \{(m, n) \in \mathcal{N} \times \mathcal{N} \mid n^2 = m\}$

 (g) $R = \{(m, n) \in \mathcal{Z} \times \mathcal{Z} \mid n^2 = m\}$

(2) Give the domain and range for each of the following functions.

 (a) $\{(x, y) \in \mathcal{R} \times \mathcal{R} \mid y = x^2 - 5\}$

 (b) $\{(x, y) \in \mathcal{R} \times \mathcal{R} \mid y = \tan x\}$

 (c) $\{(x, y) \in \mathcal{R} \times \mathcal{R} \mid y = 1/(x - 1)\}$

 (d) $\{(x, y) \in \mathcal{R} \times \mathcal{R} \mid y = \chi_{\mathcal{N}}(x)\}$

 (e) $\{(x, y) \in \mathcal{R} \times \mathcal{R} \mid y = 0\}$

 (f) $\{(x, y) \in \mathcal{R} \times \mathcal{R} \mid y = \sqrt{x^2}\}$

(3) Prove that the following relations are not functions on \mathcal{R}.

 (a) $\{(x, y) \in \mathcal{R} \times \mathcal{R} \mid x^2 = y^2\}$

 (b) $\{(x, y) \in \mathcal{R} \times \mathcal{R} \mid x^2 + y^2 = 4\}$

 (c) $\{(x, y) \in \mathcal{R} \times \mathcal{R} \mid x = \sin y\}$

 (d) $\{(x, y) \in \mathcal{R} \times \mathcal{R} \mid y = \sqrt{x}\}$

(4) Let $f \subseteq A \times B$ and $g \subseteq C \times D$ be functions. Prove that $f \cap g$ is a function from $A \cap C$ to $C \cap D$.

(5) Let $f \subseteq A \times B$ and $g \subseteq C \times D$ be functions. Show that if $A \cap C = \emptyset$, then $f \cup g$ is a function from $A \cup C$ to $B \cup D$.
Give an example to show that the assumption $A \cap C = \emptyset$ here is essential.

(6) Regarding the bijection $f : A \to B$ as a subset of $A \times B$, write the definition of f^{-1}.

(7) Regarding the function $f : A \to B$ as a subset of $A \times B$, write the definitions of f being (a) one-one, and (b) onto.

(8) Regarding the functions $f : A \to B$ and $g : B \to C$ as subsets of $A \times B$ and $B \times C$, respectively, write the definition of the composition function $g \circ f$ as a set of ordered pairs.

5.7 An Example: The Reals

In this book we assume that you already have a reasonable idea of the concept of a real number. But why do we need to introduce the real number system in the first place? In real life, we are never able to measure any quantity (e.g., length or temperature) to more than a few decimal places, for which the rational numbers suffice. Indeed, the rational numbers suffice for the measurement of any quantity to whatever degree of accuracy is required. This fact is illustrated by the following simple result:

THEOREM 5.7.1
If r, s are rationals, $r < s$, then there is a rational t such that $r < t < s$.

PROOF Let

$$t = \frac{1}{2}(r + s)$$

Clearly, $r < t < s$. But is $t \in \mathcal{Q}$? Well, letting $r = m/n$, $s = p/q$, where $m, n, p, q \in \mathcal{Z}$, we have

$$t = \frac{1}{2}\left(\frac{m}{n} + \frac{p}{q}\right) = \frac{mq + np}{2nq}$$

so as $mq + np$, $2nq \in \mathcal{Z}$, we conclude that $t \in \mathcal{Q}$. ⬜

The above property of there being a third rational between any two unequal rationals is known as *density*.

The trouble is, although the rationals are dense (in the above sense), there are nevertheless "holes" in the rational line. For example, if we let

$$A = \{x \in \mathcal{Q} \mid x > 0 \wedge x^2 < 2\}$$
$$B = \{x \in \mathcal{Q} \mid x > 0 \wedge x^2 \geq 2\}$$

then every element of A is smaller than every element of B, and

$$A \cup B = \mathcal{Q}$$

But A has no greatest member and B has no smallest member (as you can easily check), so there is a sort of hole between A and B. This is the hole where $\sqrt{2}$ ought to be, of course. The fact that \mathcal{Q} contains holes of this nature makes it unsuitable for some mathematical purposes, even though it suffices for all our measurements. Indeed, a number system in which the equation

$$x^2 - 2 = 0$$

has no solution is a pretty weak one.

Suppose we have agreed that for mathematics we require a number system which does not contain such holes and in which we can take square roots, etc. (at least for positive numbers). How do we go about finding such a system? Or, to put the question another way (since everyone reading this book will know about the real number system), what exactly *are* the real numbers?

If we ask what the rationals are, we can answer by referring back to the integers, since all rationals are essentially quotients of integers. (More specifically, they can be defined as equivalence classes of pairs of integers. See Exercise 5.5 (4) to get some indication of the general idea.) Since the integers are so basic, for most purposes this suffices as an answer.

But when the same question is asked of the reals the answer is not so easy. The point is, although the rationals are an abstract system of entities, and thus nothing more than figments of our intuition, they correspond to such basic concepts that they possess a certain concreteness. (We can calculate a number to 30 decimal places, even 50, and to get all the rationals we just need to imagine that we could calculate to *any* finite number of places, which does not require too great a feat of imagination.) But the concept of an irrational number is already much more abstract. As we proved in Chapter 2, it is utterly impossible to express $\sqrt{2}$ as a quotient of two integers; the decimal expansion of $\sqrt{2}$ continues forever without recurrence. So in what sense are we justified in regarding, say, $\sqrt{2}$ as a number at all? The answer is, only in a strict mathematical sense. One constructs the real number line by an abstract mathematical construction which, essentially, "fills in all the holes" in \mathcal{Q}.

This construction is quite complicated, and was one of the major mathematical accomplishments of the 19th century. It only answers the question "What is a real number?" in an abstract, mathematical sense. Its value is that it does tell us that we are not going to run into any trouble if we imagine that the holes in \mathcal{Q} are all filled in. And the advantage of filling in the holes is that the mathematical possibilities

multiply enormously. (For instance, the calculus is much easier to develop over \mathcal{R} than over \mathcal{Q}.)

For our present purposes, we shall take as a sort of definition of a real number the points on the real line. This is, of course, a circular definition, and hence not really a definition at all, but it does provide us with a useful and intuitive starting point.

The set of reals (i.e., the points on the real line) is denoted by \mathcal{R}. The elements of \mathcal{R} possess a natural ordering ($<$) which corresponds to the ordering of the points on the real line.

The fact that the real line has no holes is expressed by the so-called *completeness property*, described below.

5.8 Upper Bounds. Completeness

Let $A \subseteq \mathcal{R}$. Any real x such that

$$(\forall a \in A)(a \leq x)$$

is called an *upper bound* of A. We call x a *least upper bound* (abbreviated *lub*) if there is no smaller upper bound: that is to say, whenever $y < x$, y fails to be an upper bound of A. Clearly then, x is a least upper bound of A iff:

- $(\forall a \in A)(a \leq x)$; and

- whenever $y < x$ there is an $a \in A$ such that $a > y$.

A set A of real numbers does not have to have any upper bound. For instance, the set $\mathcal{N} \subseteq \mathcal{R}$ has no upper bound. But the completeness property for the real line says that if A does have an upper bound, it will necessarily have a least one.

Completeness property
Let $A \subseteq \mathcal{R}, A \neq \emptyset$. If A has an upper bound, then it has a least upper bound (in \mathcal{R}).

If A has a least upper bound, this must be unique. For if x and y were two distinct least upper bounds, one would have to be smaller than the other, say $x < y$, and then y would not be a *least* upper bound.

The rational line, \mathcal{Q}, does not have the completeness property. For instance, the set

$$A = \{r \in \mathcal{Q} \mid r^2 < 2\}$$

is bounded above in \mathcal{Q} by 2. But it has no least upper bound in \mathcal{Q}. To see this, let $x \in \mathcal{Q}$ be any upper bound of A. We show that there is a smaller one (in \mathcal{Q}).

Let $x = p/q$, where $p, q \in \mathcal{N}$.

Suppose first that $x^2 < 2$. Thus $2q^2 > p^2$. Now, as n gets large, the expression $n^2/(2n+1)$ increases without bound, so we can pick $n \in \mathcal{N}$ so large that

$$\frac{n^2}{2n+1} > \frac{p^2}{2q^2 - p^2}$$

Rearranging, this gives

$$2n^2q^2 > (n+1)^2 p^2$$

Hence

$$\left(\frac{n+1}{n}\right)^2 \frac{p^2}{q^2} < 2$$

Let

$$y = \left(\frac{n+1}{n}\right)\frac{p}{q}$$

Thus $y^2 < 2$. Now, since $(n+1)/n > 1$, we have $x < y$. But y is rational and we have just seen that $y^2 < 2$, so $y \in A$. This contradicts the fact that x is an upper bound for A.

It follows that we must have $x^2 \geq 2$. Since there is no rational whose square is equal to 2, this means that $x^2 > 2$. Thus $p^2 > 2q^2$, and we can pick $n \in \mathcal{N}$ so large now that

$$\frac{n^2}{2n+1} > \frac{2q^2}{p^2 - 2q^2}$$

which becomes, upon rearranging,

$$p^2n^2 > 2q^2(n+1)^2$$

i.e.,

$$\frac{p^2}{q^2}\left(\frac{n}{n+1}\right)^2 > 2$$

Let

$$y = \left(\frac{n}{n+1}\right)\frac{p}{q}$$

Then $y^2 > 2$. Since $n/(n+1) < 1$, $y < x$. But for any $a \in A$, $a^2 < 2 < y^2$, so $a < y$. Thus y is an upper bound of A less than x, as we set out to prove.

On the other hand, the set A (regarded as a subset of \mathcal{R} now) does have a least upper bound in \mathcal{R}, namely $\sqrt{2}$.

Exercises 5.8

(1) Let $A \subseteq \mathcal{R}$, $x \in \mathcal{R}$. Show that x is the lub of A iff:

 (a) $(\forall a \in A)(a \leq x)$; and

 (b) $(\forall \epsilon > 0)(\exists a \in A)(a > x - \epsilon)$.

(2) By negating (a) and (b) above, obtain a symbolic definition of the property "x is not the lub of A." (Put your answer in a positive form.)

(3) What is lub (a, b)? What is lub $[a, b]$? What is $\max(a, b)$? What is $\max[a, b]$?

(4) Let $A = \{|x - y| \mid x, y \in (a, b)\}$. Prove that A has an upper bound. What is lub A?

(5) In addition to the completeness property, the *Archimedean property* is an important fundamental property of \mathcal{R}. What this says is that if $x, y \in \mathcal{R}$ and $x, y > 0$, there is an $n \in \mathcal{N}$ such that $nx > y$.

 Use the Archimedean property to show that if $r, s \in \mathcal{R}$ and $r < s$, there is a $q \in \mathcal{Q}$ such that $r < q < s$. (Hint: pick $n \in \mathcal{N}$, $n > 1/(s - r)$, and find an $m \in \mathcal{N}$ such that $r < (m/n) < s$.)

(6) (a) Define the notions *lower bound, greatest lower bound (glb)* of a set $A \subseteq \mathcal{R}$.

 (b) Give a characterization of the glb of a set A analogous to the one for lub given at the start of this section.

 (c) Prove that the Completeness Property is equivalent to the following proposition: if a set $A \subseteq \mathcal{R}$ has a lower bound, it has a greatest lower bound.

 (d) Let $A = \{r \in \mathcal{Q} \mid r > 0 \wedge r^2 > 3\}$. Show that A has a lower bound in \mathcal{Q} but no greatest lower bound in \mathcal{Q}.

5.9 Sequences

Suppose we associate with each natural number n a real number a_n. The set of all these numbers a_n, arranged according to the index n, is called a *sequence*. We denote this sequence by

$$\{a_n\}_{n=1}^{\infty}$$

Thus, the symbol $\{a_n\}_{n=1}^{\infty}$ represents the sequence

$$a_1, a_2, a_3, \ldots, a_n, \ldots$$

For example, the members of \mathcal{N} themselves constitute a sequence when assigned their usual order

$$1, 2, 3, \ldots, n, \ldots$$

This sequence would be denoted by $\{n\}_{n=1}^{\infty}$ (because $a_n = n$ for each n).

Or we could order the elements of \mathcal{N} in a different manner to obtain the sequence

$$2, 1, 4, 3, 6, 5, 8, 7, \ldots$$

This is quite a different sequence from the sequence $\{n\}_{n=1}^{\infty}$, since the ordering in which the members of the sequence appear is important. Or, if we allow repetitions we get a completely new sequence

$$1, 1, 2, 2, 3, 3, 4, 4, 4, 5, 6, 7, 8, 8, \ldots$$

(There does not need to be a nice rule involved: it may be impossible to find a formula to describe a_n in terms of n.)

Again, we can have a constant sequence

$$\{\pi\}_{n=1}^{\infty} = \pi, \pi, \pi, \pi, \pi, \ldots, \pi, \ldots$$

or an alternating (in sign) sequence

$$\{(-1)^{n+1}\}_{n=1}^{\infty} = +1, -1, +1, -1, +1, -1, \ldots$$

In short, there is no restriction on what the members of a sequence $\{a_n\}_{n=1}^{\infty}$ may be, except that they be real numbers.

Now, some sequences have a rather special property. As we go along the sequence, the numbers in the sequence get arbitrarily closer and closer to some fixed number; for instance the members of the sequence

$$\left\{\frac{1}{n}\right\}_{n=1}^{\infty} = 1, \frac{1}{2}, \frac{1}{3}, \frac{1}{4}, \ldots, \frac{1}{n}, \ldots$$

get arbitrarily closer and closer to 0 as n gets larger, and the members of the sequence

$$\left\{1 + \frac{1}{2^n}\right\}_{n=1}^{\infty} = 1\frac{1}{2}, 1\frac{1}{4}, 1\frac{1}{8}, 1\frac{1}{16}, \ldots$$

get arbitrarily closer and closer to 1. Again, the members of the sequence

$$3, 3.1, 3.14, 3.141, 3.1415, 3.14159, 3.141592, 3.1415926, \ldots$$

get arbitrarily closer and closer to π, although this example is not as good as some of the others, since we have not given a general rule for the nth term in the sequence.

If the members of the sequence $\{a_n\}_{n=1}^{\infty}$ get arbitrarily closer and closer to some fixed number a in this manner, we say that the sequence $\{a_n\}_{n=1}^{\infty}$ *tends to the limit* a, and write

$$a_n \to a \quad \text{as} \quad n \to \infty$$

So far, this is all very intuitive. Let us see if we can obtain a precise definition of what it means to write "$a_n \to a$ as $n \to \infty$".

Well, to say that a_n gets arbitrarily closer and closer to a is to say that the difference $|a_n - a|$ gets arbitrarily closer and closer to 0. This is the same as saying that whenever ϵ is a positive real number, the difference $|a_n - a|$ is eventually less than ϵ. This leads to the following formal definition:

$$a_n \to a \text{ as } n \to \infty \text{ iff } (\forall \epsilon > 0)(\exists n \in \mathcal{N})(\forall m \geq n)(|a_m - a| < \epsilon)$$

This looks quite complicated. Let us try to analyze it. Consider the part

$$(\exists n \in \mathcal{N})(\forall m \geq n)(|a_m - a| < \epsilon)$$

This says that there is an n such that for all m greater than or equal to n, the distance from a_m to a is less than ϵ. In other words, there is an n such that all terms in the sequence $\{a_n\}_{n=1}^\infty$ beyond a_n lie within the distance ϵ of a. We can express this concisely by saying that the terms in the sequence $\{a_n\}_{n=1}^\infty$ are eventually all within the distance ϵ from a.

Thus, the statement

$$(\forall \epsilon > 0)(\exists n \in \mathcal{N})(\forall m \geq n)(|a_m - a| < \epsilon)$$

says that for every $\epsilon > 0$, the members of the sequence $\{a_n\}_{n=1}^\infty$ are eventually all within the distance ϵ from a. This is the formal definition of the intuitive notion of "a_n gets arbitrarily closer and closer to a".

Let us consider a numerical example. Consider the sequence $\{1/n\}_{n=1}^\infty$. On an intuitive level, we know that $1/n \to 0$ as $n \to \infty$. We shall see how the formal definition works for this sequence. We must prove that

$$(\forall \epsilon > 0)(\exists n \in \mathcal{N})(\forall m \geq n) \left(\left| \frac{1}{m} - 0 \right| < \epsilon \right)$$

This simplifies at once to

$$(\forall \epsilon > 0)(\exists n \in \mathcal{N})(\forall m \geq n) \left(\frac{1}{m} < \epsilon \right)$$

To prove that this is a true assertion, let $\epsilon > 0$ be arbitrary. We must find an n such that

$$m \geq n \to \frac{1}{m} < \epsilon$$

Pick n large enough so that $n > 1/\epsilon$. (This uses the Archimedean property of \mathcal{R} discussed in Exercises 5.8.) If now $m \geq n$ then

$$\frac{1}{m} \leq \frac{1}{n} < \epsilon$$

In other words,

$$(\forall m \geq n) \left(\frac{1}{m} < \epsilon \right)$$

as required.

One point to notice here is that our choice of n depended upon the value of ϵ. The smaller ϵ is, the greater our n needs to be.

Another example is the sequence $\{n/(n+1)\}_{n=1}^{\infty}$, i.e.,

$$\frac{1}{2}, \frac{2}{3}, \frac{3}{4}, \frac{4}{5}, \ldots$$

We prove that $n/(n+1) \to 1$ as $n \to \infty$. Let $\epsilon > 0$ be given. We must find an $n \in \mathcal{N}$ such that for all $m \geq n$

$$\left| \frac{m}{m+1} - 1 \right| < \epsilon$$

Pick n so large that $n > 1/\epsilon$. Then, for $m \geq n$,

$$\left| \frac{m}{m+1} - 1 \right| = \left| \frac{-1}{m+1} \right| = \frac{1}{m+1} < \frac{1}{m} \leq \frac{1}{n} < \epsilon$$

as required.

Exercises 5.9

(1) Formulate both in symbols and in words what it means to say that $a_n \nrightarrow a$ as $n \to \infty$.

(2) Prove that $(n/(n+1))^2 \to 1$ as $n \to \infty$.

(3) Prove that $1/n^2 \to 0$ as $n \to \infty$.

(4) Prove that $1/2^n \to 0$ as $n \to \infty$.

(5) We say a sequence $\{a_n\}_{n=1}^{\infty}$ *tends to infinity* if, as n increases, a_n increases without bound. For instance, the sequence $\{n\}_{n=1}^{\infty}$ tends to infinity, as does the sequence $\{2^n\}_{n=1}^{\infty}$. Formulate a precise definition of this notion, and prove that both of these examples fulfill the definition.

(6) Let $\{a_n\}_{n=1}^{\infty}$ be an increasing sequence (i.e., $a_n < a_{n+1}$ for each n). Suppose that $a_n \to a$ as $n \to \infty$. Prove that $a = \mathrm{lub}\{a_1, a_2, a_3, \ldots\}_{n=1}^{\infty}$.

(7) Prove that if $\{a_n\}_{n=1}^{\infty}$ is increasing and bounded above, then it tends to a limit.

No Answers to the Exercises

This is where you would have found the answers to the exercises — if I had included any. Here is why I have not.

I am sure that, hitherto, all of your math textbooks have had answers in the back of the book. Many current transition books that compete with mine also provide the student with a selection of answers to the exercises. And over the 20 years my book has been in print, many students, some instructors, and even the occasional editor, have suggested that I too should include such a section. On each occasion, after thinking about it at some length (and yes, I know all the arguments in favor of an answer section), I decided against it. I have two reasons for doing this.

First, the book is not written to be read in isolation. It is written for university students who are taking a course that makes the transition from calculus to modern pure mathematics — often known as a "bridge course." Attempting to answer the problems I include is intended to assist in that process. Knowing whether the answer to a particular problem is correct, however, is of relatively minor importance. Indeed, it may be misleading to know that a particular answer is "correct." What is at issue with this material is deep *understanding*. To achieve that, you will inevitably have to discuss any difficulties with your instructor. I want to force you to do just that.

My second reason is this. After completing this book (and the course it accompanies), you are supposed to be ready to enter the world of real, present-day, pure mathematics — prepared to start working in that discipline. But for the working mathematician, there is no answer in the back of the book — indeed, "the book" may not yet have been written. Few advanced-level books provide answers in the back (or colored highlighting of important points for that matter, something else I have consistently avoided). To my mind, the sooner you learn to live with that fact, the better (for you). I want you to emerge from working through this book reasonably well prepared to learn and do some interesting mathematics. Better then that the process of weaning you off from the prepackaged material of high school and first-year university mathematics courses takes place now rather than later, when the going will be considerably tougher. (Although as you are working through this book you will probably feel it can hardly get any tougher!)

Of course, in addition to approaching your instructor, there is nothing to prevent you seeking ideas, and even answers, in other books, or by asking a colleague (in your case a fellow student). Professional mathematicians do that all the time. So should anyone working through this book. But then it will be *you* who has found the result; you won't have been given it on a plate by *me*.

List of Symbols

Index